Honeybee

Lessons from an Accidental Beekeeper

Honeybee

Lessons from an Accidental Beekeeper

C. MARINA MARCHESE

Illustrations by Elara Tanguy

BLACK DOG
& LEVENTHAL
PUBLISHERS
NEW YORK

Published by
Black Dog & Leventhal Publishers, Inc.
151 West 19th Street
New York, NY 10011

Distributed by
Workman Publishing Company
225 Varick Street
New York, NY 10014

Manufactured in the United States of America

Cover and interior design by Susi Oberhelman

All illustrations © Elara Tanguy, except for pages 5 and 37,
which are drawn by the author.

Cover art from Grand Herbier by Pedanius Dioscorides,
Getty/Bridgeman Art Library

ISBN-13: 978-1-57912-815-9

h g f e d c b a

Library of Congress Cataloging-in-Publication Data available upon request.

For my honey
V-bee

Contents

Acknowledgments

Gratitude, praise, and recognition to the following people I have been lucky to learn from and work with on this personal journey:

My agent, Coleen O'Shea, and her business partner, Marilyn Allen, both of whom were *bee-lievers* in this project from the beginning. I am indebted to author Claire Garcia, for her enthusiasm and generous introductions, and to Connie Pappas, for sparking my motivation.

Becky Koh, editor extraordinaire, who fearlessly accompanied me into the hive and also extracted 20 pounds of sticky honey. Her brilliant guidance, editing, and vision shaped my concept into perfection. Thank you!

J.P. Leventhal, who embraced my passion for the honeybee, and everyone at Black Dog & Leventhal who contributed to making this project a success, including Liz Hartman, director of marketing and publicity; Judy Courtade, sales director; and True Sims, production manager. As well as Susi Oberhelman, for her lovely book design, and Elara Tanguy, for her charming illustrations.

Howland Blackiston, who introduced me to the magical world of honeybees. Your mentoring and creative spirit were the inspiration for this work. *Grazie mille!*

The community of amazing beekeepers I have met on my journeys at home and around the world, who have opened up their own hives and graciously shared their honeybee wisdom. The Back Yard Beekeepers Association and its members, for allowing me into their hive and enriching my journey. The Apitherapy Society, for introducing me to how the honeybee heals. Kim Flottum at *Bee Culture* magazine, who gave me the confidence to write. Master beekeeper, Ann Harmon, and Alan Lorenzo, bee venom therapist, who read and advised. Honey judges Robert Brewer and Michael Young, both of whom inspired my reverence for honey. Bill and Royal Draper, for answering all my bee questions. Giovanni and Francesco at Bottega della Api, *Siete molto gentili*. Joe at Puglia Wine Imports.

Vic, who patiently read, forfeited vacations, and put up with my deadlines, yet has not swarmed and continues to keep honeybees. My sister Nicole, for reading. My sister Andrea, for bottling honey. Mom, for her honey recipes, and Dad, who is now a true honey lover. Sarah, my tall, clear glass of water. And, *Brave!* to all the worker bees at Red Bee Honey who kept the hive buzzing so I could write.

Special beekeepers, chefs, and loyal honey lovers who gave their generous support along the way. Thank you, Nick, Taylor, and Amy at Murray's Cheese; Erin at Artisanal Premium Cheese Center; Julian and Lisa Niccolini at The Four Seasons; and Marty Vaz at Speak Easy Cocktails.

My Sweet Encounter with the Honeybee

Although I've been a beekeeper for a long time, I will never forget my very first taste of fresh honey straight out of the beehive. Almost ten years ago a neighbor, Mr. B, invited me to his apiary to meet his honeybees. I was apprehensive about the offer. I thought to myself, "Sure, I like honey, but I'm not so sure I like honeybees." Suddenly I imagined myself surrounded by a swarm of hundreds of buzzing bees. The idea scared me, as I think it would most people. But I was ready for a new adventure, so I accepted Mr. B's invitation.

It was a perfect early spring day when I showed up at Mr. B's home to meet his honeybees. In his backyard stood three tall boxes that looked like painted white file cabinets; these were his beehives. As Mr. B greeted me, he handed me a beekeeper's veil to put over my head for protection. Then he donned his own veil and walked toward the hives. As I followed him, heart pounding in my ears, he explained that honeybees, although quite docile, were also curious creatures. They liked to crawl into nooks and crannies and into our clothing. The veils should stop them from stinging our faces. "Stinging our faces?" I wondered what I was getting myself into. By the time we arrived at the hives, I was trembling. Mr. B lit his bee smoker, a small tin container that looked a little like a coffee can, and blew a few puffs of smoke into the front entrance of the first beehive. Then he lifted the cover to direct the smoke at the bees inside. He explained that the smoke calmed the bees and distracted them from our presence.

He then gently removed the cover completely from the hive and placed it on the grass. I craned my neck to peer inside, still trying not to get too close. Hundreds, maybe thousands, of honeybees crawled across the top of the ten perfectly positioned wooden frames that sat vertically inside the box. I was utterly surprised and relieved to see that the bees were indeed quite calm. With his bare hands, Mr. B. slowly removed a single wooden frame covered with bees. I watched with amazement as the bees walked across his fingers, then his hands, and onto his sleeve. But Mr. B took no notice. "These are *Italian* honeybees," he said. I had to smile. Since I am of Italian ancestry, I liked the idea of Italian honeybees. Out of nowhere came thoughts of telling my friends, "I raise *Italian* honeybees."

Mr. B inspected the frame and pointed out the different kinds of bees: the female worker bees that gathered the nectar and made the honey, and the male drone bees whose primary job was to mate with the queen. He told me that there was one queen bee in every hive and that all hive activities revolved around her egg-laying schedule. The female ruled the hive—I liked the way that sounded.

When Mr. B announced that it was my turn to hold the frame, I shrank back. But his gentle handling of the bees and his calm demeanor somehow gave me the courage to accept the frame from him with my own bare hands. Bees were everywhere—dozens of them crawling on my fingers and making their way onto my sleeves. I took a deep breath and held the frame firmly so as not to make any sudden movements and upset them. "I can do this bee thing," I said to myself. "I am fearless."

As I held the frame, Mr. B pointed out the perfectly formed honeycomb, made of beeswax, that filled the center of the frame. The honeycomb was where the queen laid her eggs and the worker bees stored their pollen and honey. When I held the frame up to the sunlight, the honeycomb looked like a beautiful stained-glass window. Mr. B. poked his finger into the hexagon-shaped cells. Sparkling amber liquid oozed out of the cells and drizzled down the frame. Mr. B stuck his fingers under his veil and carefully licked off the precious honey. He invited me to do the same. Careful not to disturb a single bee, I poked my finger into a new cell to expose more of the pristine honey. As I excitedly drew my finger up to my mouth, I forgot about my protective veil and smeared it with the honey. Mr. B chuckled. I captured another dollop of honey, this time managing to bring my finger underneath my veil. It tasted glorious and exquisite, heavenly and perfect. It was like nothing I had ever savored. At that moment, I knew I wanted to keep Italian honeybees that made this divine treasure called honey.

I took home a bottle of Mr. B's pure honey, and I proudly put it on my kitchen windowsill. Each morning after that, I woke to see the sun shining through the amber bottle, beckoning me. I began to use the honey instead of sugar in my espresso, and it also soon became a decadent spread for toast. I mixed that honey into my salad dressings at lunch and marinades for chicken at dinner, and I swirled it on my ice cream with walnuts for dessert. Over the course of a week, using the honey became a hedonistic ritual, and the windowsill was the honey's altar. I guarded that treasured bottle and offered nibbles only to those who were worthy of its sensuousness. Crazy for this thing called honey, I became obsessed with the countless ways to enjoy it. A favorite became bergamot iced tea flavored with mint from my garden and a few heaping spoonfuls of honey.

At one point during the week, a crazy thought occurred to me: Is it safe to eat honey right out of the beehive? Is a beehive clean?

Bergamot Iced Tea with Red Bee Honey

SERVINGS: 4

INGREDIENTS:

4 cups boiling water
2 Earl Grey tea bags
¼ cup Red Bee® wildflower honey
½ cup fresh mint leaves

Bring water to a boil. Steep tea bags in water for three minutes. Dissolve honey, then add mint while warm. Serve over ice cubes when cool.

Mr. B and I had just opened up the hive and stuck our fingers in there. Didn't the honey have to be processed or pasteurized? A few days later, when I spoke to Mr. B again, he assured me that raw honey cannot carry bacteria because of its low pH. I could not think of any other food with that quality. He also told me that the bees always keep their hive pristine. He described how he extracted honey from those luscious honeycombed frames, separated it from the wax by straining it through a cheesecloth or pair of women's nylon pantyhose, and then simply poured it into bottles. I thought, "Hmm, honey, straight from the hive and into the bottle—no boiling, no sterilizing, no refrigeration, no nothing. Could making and bottling honey be that easy?"

My first visit to Mr. B's apiary would change the course of my life. In time, I would become immersed in beekeeping and honey. I would discover that there were bee meetings all over the country and even the world. It wasn't long before I would comb gourmet food shops in search of unique honeys, and each purchase would symbolize a completely different culture and culinary experience. Beekeeping would give me a new perspective on food and my interaction with nature. I would begin introducing honey into my daily diet, using it to replace processed sugar and artificial sweeteners. My general health would improve markedly. Honey even helped me beat colds and then alleviate my allergies. The notion of the queen ruling her hive and female worker bees gathering nectar and making honey appealed to my own sense of industry and spurred me to launch my own honey business. There has been no end to the wisdom the honeybees have given me. These tiny creatures are symbols of craftsmanship, dedication, and perfection, and for all of their lessons I am eternally grateful.

Honeybee

CHAPTER 1

My Life as a Worker Bee

The Monday after my first visit with the honeybees, I was back commuting from my home in Connecticut to my office in New York City. I was creative director for a small giftware company, developing gifts and home-accessory products to be manufactured overseas, then brought back to the United States to be sold at trade shows and at our retail shop in Greenwich Village. The best part of my job was researching and shopping for new ideas. I was given considerable creative freedom and the luxury to travel to China, which is how I survived the more mundane aspects of my day-to-day work. Still, many times the owners of the company shot down my best concepts because they thought the ideas were too outlandish or would not lead to enough sales to make it worth the manufacturing efforts. Products always followed trends and it was my job to adhere to them, but not too closely. Frequently my designs needed to be watered down and made palatable for the general public. Our biggest challenge as a small company was competing with the larger stores that could make more products faster and cheaper. It could be an exhausting process— one that could quickly and easily dull the creative spirit. And I'd been feeling very weary of it.

As I stood on the platform in the early morning sunshine, waiting for my train, I realized that where I really wanted to be was back in the bee-yard. The exhilarating experience of communing with those bees was still fresh in my mind. Suddenly, I craved the country life, to be in the garden, toiling under the sun, caring for my very own Italian honeybees. "What was really standing in my way?" I wondered. Of course, the obligation of my work often took me away from home for several weeks at a time. Who would care for my bees while I was gone? Would they starve, or miss me? Did they need special attention? Clearly, I needed to learn more.

That morning, the train was late, so I wandered inside the station house. This particular station house had no seats, but did have a free, communal book rack where commuters could take and leave books as they passed through the station. The books gave the station a cozy feeling. I perused the rack as I waited, and a red book with the word "beekeeper" on the cover caught my eye. To my utter delight, I saw it was titled *The Beekeeper's Apprentice*. I plucked the old, tattered paperback off the shelf to examine it. It turned out to be a mystery about Sherlock Holmes, who, I was elated to find, kept honeybees. The book was just the remedy I needed, a small distraction from the workday that lay ahead.

Once I got onto the train, I dove right into that book. A huge Sherlock Holmes fan, I was fascinated and charmed by the story, in which Holmes meets a young intellectual woman studying bee behavior. They partner together for adventures in sleuthing and beekeeping. Life was imitating art: I was like the young apprentice to my own Holmes, Mr. B. Was this story a sign of what my future held? My destiny was calling, and I was beginning to feel I had to respond.

• • •

FOR THE NEXT SEVERAL WEEKS, I continued my regular commute back and forth to the city, working on new product designs and selling them at the store until it was time for one of my regular trips to China.

At this point I had been to China twice a year for almost six years to source out and oversee manufacturing of our company's giftware designs. I was working at a factory in a small, remote city outside of Huangzhou, approving the final details of some new products. It was a sweltering day in China; the humidity was sky-high, and the air dense. My associate Mr. Wang offered to take me to his favorite noodle house for lunch. On the way, I noticed a few ragged handmade tents alongside the dirt road. In front of one tent stood a dozen beehives. The streets of China are always full of unexpected surprises, and if I had not still had bees on the brain, I might have walked right by without giving them a second thought. Instead, I stopped dead in my tracks and stared as though I had discovered the eighth wonder of the world.

"Mr. Wang, those are honeybees hives!"

Mr. Wang was puzzled by my excitement. As we drew closer, I could hear the low hum of bees at work. I asked Mr. Wang if it was okay for me to watch for a few minutes. The wooden beehives were similar to the ones in Mr. B's backyard. Tools and equipment of all kinds were piled up beside the hives. I recognized a few smokers, but for the most part the items were foreign to my untrained eyes. A family of beekeepers was tending the hives. Two of the men began smoking and opening up one of the hives. Although dozens of bees were flying around them, they did not wear protective clothing or veils. One held a frame similar to the one I had held at Mr. B's apiary, except this one was perfectly clean, not covered in honey or bees. The other removed a single bee-covered frame from the hive and held it up to the sun. Shimmering in the sunlight, the frame was obviously full of luscious honey. The other beekeeper slipped his clean frame into the slot where the honey-covered frame had been. Before they closed up the hive, they brushed all the bees off the honey-filled frame and back into the top of the hive. Then they carried off the frame of honey with just a few stubborn honeybees following behind. They wrapped it with newspaper, placed it in a wooden box, and sealed up the box. A woman standing next to the tent accepted the box of honey and paid for it with cash. Mr. Wang told me that this is how many Chinese people get their honey—directly from beekeepers.

"Honey is an ancient tradition here, used for its health benefits. And honeybees are respected," he said.

Before continuing to the restaurant, I took some photographs to remember what I'd seen—and, of course, to show Mr. B.

Now that I knew there were honeybees in China, I made it my mission to purchase some local honey. Mr. Wang said there was a honey shop not too far away from the factory, and after lunch we set out for it. He told me stores that sold only honey and honey products were common in China, and his wife regularly purchased honey for their family, especially when one of the children had a cold.

Turning down a side street, we arrived at a small shop with a distinct honeybee logo on its door. The interior of the shop was decorated like a beehive. There were shelves quite cleverly designed in the familiar hexagonal shape of a honeycomb, and each displaying a single jar of honey. The honeys were different shades of amber and gold. We also found real honeybees on display inside a framed glass box; this chamber of wonders allowed viewers to peek into the inner sanctum of a real beehive. A slight hum pulsed from the framed glass box as thousands of busy honeybees crawled across the honeycomb. Mr. Wang and I could feel the heat of their little bodies permeating through the glass and smell the unmistakable aroma of honey and beeswax as they went about their business.

The clerk behind the counter was busy filling a huge glass jar with honey from a stainless steel tank. A customer watched her intently, as though his honey purchase were a ceremonious undertaking. Mindful not to spill a single drop, the clerk scooped up the honey with a primitive-looking ladle and drizzled it into the container. When the jar was full, the clerk twisted the cap tightly, wiped down the jar with a rag, and brought it to the register. After a brief conversation, the customer paid and was on his way.

Waiting my turn, I peered into the glass display counter exhibiting fine specimens of various honeys, all with beautiful labels that enticed my artistic eye. Not understanding Chinese, I relied on Mr. Wang to translate for me. There was loquat honey, million-flower honey, rose honey, and many others that he was not able to translate into English. He also pointed out jars that contained a chunk of honeycomb straight from the beehive and other jars of thick, creamy honey. I had no idea there were so many types. Could any of these types of Chinese honey taste much different from the honey back home? I needed to know so I opted for a single jar of the million-flower honey, which later I learned is more commonly called wildflower. Mr. Wang motioned the clerk over to tell her which bottle we wanted. She wrapped my honey and placed it in a bag decorated with bees.

When we arrived back at the factory, the perpetual teapot was boiling. I could not resist having a cup of the locally grown green tea and drizzling it with my newly purchased honey. It was golden amber with a slight orange tinge, and it fell off the spoon in a thick spin. The flavor was as scrumptious as Mr. B's honey, but in a different way. It carried different tasting notes that I did not yet have the vocabulary to explain. In my amateur opinion, it was exquisite and a treasure to be savored. Million-flower honey was my second jar of "pure" honey. I had officially begun my quest of collecting and hoarding honeys. I also learned something else in China: I learned that I was ready to become a beekeeper.

CHAPTER 2

Becoming a Beekeeper

The decision to start my own beehives was an unbelievably liberating moment. With the taste of Chinese honey still on my tongue, I felt courageous and empowered to be embarking on a new hobby as offbeat as beekeeping. Friends and family would certainly think I had lost my mind, and my neighbors would think I was joking. But I didn't care. I had tasted the divine honey, and I was hooked. My little red cottage on the outskirts of Weston craved a romantic-style garden where my honeybees would thrive. My garden would be a banquet of nectar and pollen, and my own honeybees would visit the fauvist flower beds to pollinate my flowers and vegetables. I relished the idea of working outside with the bees, having dirt under my finger-nails, and bonding with nature. The seductive smell of beeswax would always scent the yard. And, of course, there would be honey everywhere, every day.

Before I dove into my new hobby completely, Mr. B urged me to attend a few bee meetings of a local beekeepers club to learn more. "Bee clubs?" I thought. "Is there really such a thing?" Indeed, there is. It turns out that beekeepers, like the bees they keep, are extremely social creatures. The Back Yard Beekeepers Association is one of many hobbyist beekeeping clubs in the state of Connecticut. There are more than 350 members in the BYBA (as it is known locally), all of whom keep and love honeybees and, of course, honey. Club members are dedicated to volunteering their time and expertise to promote this ancient craft by offering many outstanding educational opportunities for those who want to begin keeping honeybees or just wish learn more about these fascinating creatures. I felt the club members' warmth and electrifying enthusiasm the minute I walked into the church hall where the meeting was held.

I was greeted at the door by an adorable beekeeping couple selling tickets for their monthly bee raffle. They were both wearing bee T-shirts; the gentleman wore a cap adorned with dozens of bee pins, and bee earrings dangled from the woman's ears. Welcoming their newest attendee, they offered me flyers about the evening's events and invited me to purchase a raffle ticket. The humble table beside them held the prize: an odd piece of bee equipment. Although I had no idea what it was or how it worked, I paid for a chance to win it. With my lucky ticket in hand, I entered into the hall.

Inside, the hall was buzzing with beekeepers. Along one side of the room was a table with food, and another table was set up with the official beekeeping library and book sale. There were books, movies, and pamphlets, and a librarian was signing out these materials to interested borrowers. I walked the length of the table, gazing at the many books—new and old, serious and lighthearted. The titles included *How to Plant a Bee Garden*, *Beekeeping for Fun and Profit*, and *Biology of Honeybees*, and the topics ranged from bee rearing to bee gardening to cooking with honey. I was beginning to realize that the topic of beekeeping stretched far and wide. I was especially attracted to the older books with artistic etchings of bees and historic photographs of primitive beehives.

I spied a beginner's book for beekeepers; it seemed accessible and uncomplicated, and deciding I would need all the help I could get, I grabbed it. After checking out my new book, I headed over to the snack table to sample the homemade snacks, all of which were made with local honey. Biscotti cookies, barbecue chips, nuts—it was a sticky spread noble enough for any queen and regally displayed upon a bee-embroidered tablecloth. I indulged in a few honey-laced cookies.

Meanwhile, I overheard authentic beekeeping chitchat.

"How are your bees doing, Tom?" asked one woman.

"Two of my colonies are doing fine. They wintered over nicely. And the other swarmed last weekend, and I never did see where they settled down."

"Yes, it's swarm season, and we lost a couple of hives as well. Now I am looking for a new queen for one of my hives."

A voice came over a loudspeaker, announcing the meeting was about to begin. A gentleman on the makeshift stage introduced himself as the president of the club. The crowd mellowed to a quiet hum as each beekeeper found a seat. I grabbed another honey cookie and sat in the back row with some handouts and my new book. As I looked around, I noticed the audience was a real mix of people from all walks of life. Whenever I had heard the word "beekeeper," I imagined an elderly man with big hands, wearing a straw hat and coveralls; Peter Fonda in *Ulee's Gold* immediately came to mind. But this group was full of men and women, gardeners and farmers, professionals and business types, artists and other creative souls.

The president announced the upcoming hive inspections and the many other bee-related educational opportunities that were open to club members. All these events and more were listed on the group's Web site. Someone in the back of the room reminded the group there would be a honey swap that evening after the speaker. *Honey swapping!* That was something I definitely wanted to see.

Honey Almond Biscotti

SERVINGS: 36 COOKIES • PREP TIME: 25 MINUTES

INGREDIENTS:

½ cup unsalted butter or margarine, softened
¾ cup Red Bee® clover honey
2 large eggs
1 teaspoon vanilla extract
3½ cups all-purpose flour
2 teaspoons anise seeds
2 teaspoons ground cinnamon
½ teaspoon baking powder
½ teaspoon salt
¼ teaspoon baking soda
¼ teaspoon dried cranberries
¾ cup dried slivered almonds

Preheat oven to 350°F.

Using electric mixture, beat butter until light; gradually add honey, eggs, and vanilla, beating until smooth. In a small bowl, combine flour, anise seeds, cinnamon, baking powder, salt, and baking soda; gradually add to honey mixture, mixing well. Stir in cranberries and almonds.

Shape dough into two 10x3x1-inch logs on greased baking sheet. Bake for 20 minutes or until light golden brown. Remove from oven to wire rack, and cool 5 minutes. Reduce oven to 300°F. Transfer logs to cutting board. Cut each log into ½-inch slices; arrange pieces on baking sheet. Bake 20 minutes or until crisp. Cool on wire racks.

Finally, the guest speaker was introduced. He was a well-respected entomologist and researcher who spoke about honeybees and the vital role they play in the pollination of our food. I listened with rapt attention throughout his entire lecture. There was so much new and interesting information to take in.

POLLINATION:
WHY WE CANNOT LIVE WITHOUT IT

Before plants can grow, they need to be pollinated. Pollination is the first necessary step in the fertilization of all plants through the transfer of *pollen*, the sticky, yellowish dust produced by a plant's flower. Honeybees are responsible for pollinating more than 100 agricultural crops in the United States, including fruits, vegetables, seeds, legumes, and sixteen types of flowers species.

Honeybees make at least twelve foraging trips in a single day, visiting several thousand flowers. Tempted by brilliant colors and pleasing scents of nectar, the honeybees are rewarded with pollen. The process of pollination begins with the tiny grains of pollen produced by the male reproductive part of a flower, which is called the *anther*. When this pollen is moved to the female part of a flower, called its *stigma*, pollination occurs. These pollen granules are transferred as they stick to the hairy bodies of worker bees during their visits to each plant. Although there are many ways pollen can be transferred—butterflies, bumblebees, moths, hummingbirds, bats, and even the wind can move pollen—honeybees are among the most efficient pollinators because their hairy bodies attract pollen and distribute it while they are foraging.

After pollination, the male pollen unites with the female egg inside the plant's ovary. There it grows, or germinates; this final step of the process is called fertilization. Once fertilization takes place in the ovary, the flower expires and drops its petals from the stem, and a fruit and a seed are produced. The seeds spread, and new flowers grow.

Some species of plants can be fertilized by pollen from flowers on the same plant or from a similar *cultivar* of the same species. Although

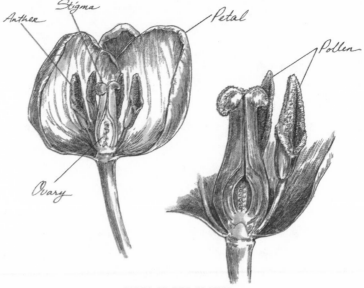

Anther Stigma Petal Pollen Ovary

PARTS OF THE FLOWER

these self-pollinated plants can provide their own pollen, they still need bees to transfer the pollen to another flower on the same plant. Other species of plants require pollen from a different cultivar for fertilization, and this process is called *cross-pollination*. Many of these flowers have male and female parts, but nature will not allow these plants to fertilize themselves with their own pollen. Cross-pollination is only possible when both plants are in bloom at the same time.

A single honeybee tends to visit the same type of flowers repeatedly while gathering pollen. Others from her same colony may visit different types of flowers, but each bee will visit its own chosen type over and over. This loyalty is called *flower constancy*, and it is another reason honeybees are such great pollinators.

Flowers also produce nectar, a clear liquid substance usually found in the center of most flowers, which is called a *nectary*. Worker bees gather nectar from flowers and carry it back to the hive to be made

Pollination Terms

Cultivar: A cultivated variety of plants of the same species given a unique name and bred for their desirable characteristics.

Variety: A variation of a species.

Self-pollination: The transfer of pollen from the male part of one flower to the female part of the same flower or other flowers on the same plant. Also called "perfect flowers."

Cross-pollination: The transfer of pollen from the male part of the flower to the female part of the flower of a different cultivar.

Pollinator: Agents such as bees, moths, butterflies, other insects, the wind, and humans that physically transfer pollen from the anther to the stigma of a flower.

Pollinizer: A plant that is a source of pollen.

into honey. Honeybees eat pollen for its protein, vitamins, and fat and honey for its carbohydrates. In return, the honeybees provide essential pollination to the plants they visit.

Crops that have not been properly pollinated are disfigured and underdeveloped. They taste as unappetizing as they look. A perfectly pollinated cucumber will grow straight and well rounded; if it is not fully pollinated, it will be lopsided and will curl. Cantaloupes that are firm, heavy, and juicy have been properly pollinated, as have apples that are large and able to stand up straight without tipping. The seeds of a watermelon tell us a lot about pollination. The black seeds were pollinated, but the white seeds were not. The more black seeds, the sweeter the watermelon will be. It takes pollinated seeds to produce the hormones that cause fruits and vegetables to ripen and taste good. Poorly pollinated crops will spoil quickly and often cause digestive troubles. Many discerning shop and restaurant owners won't accept underdeveloped produce because they know consumers will reject it.

Environmentally conscious beekeeping is a completely sustainable agricultural enterprise. Honeybees support sustainability through pollination by assisting the natural life cycle of plants, which replenishes the rich variety of flora and fauna in a given geographic region, while at the same time supporting wildlife. Meanwhile, the honey the bees make provides a natural edible product that does not harm the environment, creates no waste, and maintains the ecosystem. Honey production can be a sign of the healthy biodiversity of a geographic area. Large numbers of healthy plants indicate a sustainable environment. Choosing to eat honey from your local geographic region is a sustainable choice. By eating local honey, you are not only supporting your local beekeepers but also helping agriculture and small businesses stay afloat.

One-third of the food produced in the world—or one out of three bites of anything we humans eat—depends, to some degree, on honeybees. Healthy bees mean not only more honey but also a wide variety of food on our tables. Without honeybees, our supply of fresh food would be severely limited.

Beekeepers hire out honeybee populations to farmers around the country who rely on their honeybees to facilitate the pollination of their crops. Farmers secure this service through pollination contracts. Beekeepers are hired annually and are paid per hive. In 2008, they were paid $135 to $190 per hive. They often make more money renting out their colonies for pollination than they do selling their own honey. You can feel good about purchasing honey, because you know that the bottle represents not only all the work the honeybees put into making it but also the work they have done for agriculture, farmers, and the food we eat every day.

It's estimated that there are nearly 2.4 million colonies of honeybees in the United States, and each year two-thirds of these are trucked around the country to pollinate crops for farmers. A typical season might begin in California, where 580,000 acres of almond groves completely depend on honeybees for pollination. (California supplies 100 percent of the nation's almonds and 80 percent of the world's almonds.) Each February, migratory beekeepers load 1.5 million honeybee colonies onto trailer trucks just to pollinate the state's almond crop. This huge operation is the largest pollination event in the world, requiring more than half of all beehives in the United States! The bees are then trucked to the Northeast to work on the cherry and apple blossoms. New York State, for example, needs 30,000 hives to pollinate its apple crops. Next stop is the orange groves in Florida, followed by the cranberry bogs in New Jersey and Massachusetts. The bees will then journey to the clover fields of North Dakota and finally to Maine, which needs about 50,000 hives for its blueberries.

Honeybee pollination adds $15 billion to $20 billion a year to the agriculture output of the United States, but its value to the economy cannot be underestimated. Almond producers estimate that by 2012, 2 million beehives will be needed to pollinate the expected 800,000 acres of almonds. Other crops—like melons, avocados, watermelons, cantaloupe, all varieties of citrus fruits, and cucumbers—also require several visits by a honeybee before they can bear fruit.

Honeybees also play a very important part in our supply of beef, milk, cheese, and other dairy products. Farmers rely heavily on honeybees to pollinate a large part of the alfalfa and clover crops. Both are rich in protein and make up about one-third of the diet of cows. Cows can be fed grasses and grains, but these foods don't have as much protein. Well-fed cattle means tastier meat, milk, and cheese. Häagen-

Dazs claims that 40 percent of the ingredients in its ice cream flavors are dependent upon honeybees.

Coconuts, olives, peanuts, rape, soybeans, and sunflowers also require honeybee pollination to produce the fats and oils we require in our diets. Eighty percent of the cotton that makes up the clothes we wear, not to mention our other household items, like rugs, bedsheets, and furniture fabrics, relies upon honeybees, as cotton crops need to be pollinated, too. Breakfast cereals, nut mixes, cookies, fruit pies, and juices, not to mention ketchup and salsa, are also dependent upon bees because honeybees pollinate many of the ingredients that go into them, such as cashews, walnuts, prunes, grapefruits, peaches, cherries, cinnamon, tomatoes, and onions.

THE DWINDLING U.S. HONEYBEE POPULATION

Where do all the bees come from?

The U.S. Department of Agriculture has estimated that there are between 140,000 and 212,000 beekeepers in the United States. The majority are hobbyists with a few to several dozen hives. There are approximately 1,600 commercial beekeepers operating in the United States that produce 60 percent of the nation's honey.

The demand for honeybees is increasing each year, but populations of managed hives have been declining from 3.5 million in 1989 to 2.3 million in 2008. This is a 34 percent decrease since the 1980s, when the Varroa mite was discovered in the United States. This reddish-brown parasite attaches to the honeybee during its metamorphosis inside the cell. While feeding on bee blood, the mite transmits bee viruses that weaken colonies and cause heavy losses each year. The Varroa mite is a serious threat to honey colonies and to the livelihood of the beekeepers who manage them.

Another major problem in the U.S. honeybee community was first reported by a commercial beekeeper in October 2006. It is known as Colony Collapse Disorder (CCD). CCD is defined by a sudden disappearance of the adult bee population of a hive; in some cases only the queen, a few adult bees, and the brood remain behind. Commercial beekeepers began opening up their hives to find them empty and not a honeybee in sight. Commercial honeybee losses set beekeepers and farmers back, and the resulting financial losses have been devastating. As honeybee colonies dwindle and beekeepers cannot provide the honeybees needed to pollinate crops, farmers' food production drops; consumers, in turn, do not have as much fresh produce and other foods for their tables, and the prices of available produce increase.

Beekeepers and scientists are baffled by the causes of this new disorder. Some of the many theories regarding CCD say it was brought on by poor honeybee health and overstressed, overworked bees. Pathogens, parasites, and pesticides are also considered to be possible causes. Others say CCD losses are due to insecticides—specifically new chemicals called neonicotinoids. Germany, Italy, Slovenia, and France do not allow this product to be used, since reports link it to massive honeybee losses. The overuse of many other chemicals and antibiotics are also found to compromise the immune systems of the honeybee.

There are many ongoing efforts to breed a hygienic or nearly perfect honeybee that will have the qualities to withstand diseases, pests, and pesticides. But until those efforts are successful, the declining bee population stemming from CCD and increased cost of maintaining honeybees only add to the difficulties commercial beekeepers already face. For example, the unfair practice of shipping inferior-quality and counterfeit honey, that may not actually be pure, from countries like Canada, Argentina, Australia, and China. In 2008, major retailers and commercial users of honey purchased tons of Chinese honey for $0.20

per pound, while it cost domestic beekeepers around $1.00 per pound to make their honey. Who can compete with that?

• • •

MUCH OF THIS INFORMATION, including the details of pollination, the vital role honeybees play in U.S agriculture, and the challenges commercial beekeepers face has been imparted to me over the years by speakers at the BYBA meetings. I had no idea how much the honeybees had to offer humans and the planet. During the lecture I heard at my first meeting, I jotted down notes about pollination, along with a few doodles of flowers and honeybees. A recurring sketch was a queen bee wearing a crown and holding a scepter; her attendants glided around her and their hive. These would be the first of many drawings and notes I would include in my bee journal.

A woman to my left glanced over at my doodles. We exchanged smiles, and after the guest speaker's talk, she introduced herself as Mary. I told her I was an illustrator turned beekeeper; she said she was an accountant turned beekeeper. We both laughed at finding another soul disenchanted with her nine-to-five job. Mary told me she had been keeping bees for three years and still considered herself a beginner. She also told me how honeybees offered her a spiritual pastime and that beekeepers were a passionate and knowledgeable group who are always willing to lend a hand and offer advice based on their own experience. This goodwill was contagious, and Mary now found herself offering her bee wisdom as a mentor to newer beekeepers. "How generous and kind," I thought. In a world of takers, here was a place where some people gave selflessly. It seemed that nature and honeybees had this effect on humans. I got the feeling that there was a true sense of sharing within this community of like-minded people.

I asked Mary where the honey swap took place and was directed to an old wooden table, were seven beekeepers hovered, eyeballing

Honey Bees

Honey

Bee Eggs Shop

farmers market

8-11 Coley Town historical society

Api Gelati Lip Balm

BEE

Bee Shop

honey jars of various shapes and sizes. The colors of the honeys varied as well, from light to medium to dark. Some had a tinge of red, and others green. Each honey was harvested from a local beekeeper who was a member of the club. The jar labels were handmade and whimsical and advertised the name of the beekeeper's apiary or farm. Many displayed the typical honeybees-and-flower motif, while others featured beehives. I chuckled at one that read: "One Pound Golden Hoard Honey. Our bees forage exclusively in backcountry Greenwich. They do not visit Stamford, Armonk, Port Chester or other such places." Beekeepers were certainly creative, clever, and definitely not lacking a sense of humor.

The rules of the swap were simple. Any beekeeper who contributed a bottle of honey was allowed to choose a bottle in exchange. I watched as each bottle was snatched up and wished I had some honey of my own to share. Mary, who had brought a jar, chose a very dark-colored honey called buckwheat. She told me that darker honeys had loads of iron and antioxidants. Also, the honey would have qualities of the plant whose flowers the bees took nectar from—in this case, buckwheat. One gentleman beekeeper noticed my interest in all the honey samples and offered me one of the three bottles he'd taken from the table. I accepted it with great pleasure.

"You're new here, aren't you?" he said. When I replied that I was, he said, "You keep coming here, and you'll learn a lot about bees. And make some new friends."

Before the night was over, I ran into Mr. B, who introduced me to even more beekeeping friends. "New-bee" is the name used to describe beginner beekeepers, and that's how I was referred to. I was invited to sign up as a club member and to attend a hive inspection that coming Saturday. Without hesitation, I accepted the invitation and wrote down the address.

My first beekeeping meeting was like opening a door into a whole new world that I had no idea existed. Looking back, the bee club was like a secret society, complete with its own meetings and even its own vocabulary. Through the BYBA, I would meet people from all walks of life, but who all shared a passion for bees. They would teach me how to manage my honeybees and how to harvest and bottle my honey. And they would be there for me as lifelong mentors, making me a better beekeeper. I went to bee meetings, including some in other cities, for my entire first year of beekeeping. It was not long before I became more involved in the organization, but it wasn't until after I began serving on the board of directors that I learned my Mr. B was a past president.

CHAPTER 3

Apiology: The Study of Bees

Knowing that the hive inspection awaited me on Saturday, I don't know how I managed to get through the rest of the workweek after that first meeting of the Back Yard Beekeepers. Despite all the creative freedom my job offered me, the repetitive designing and sourcing of products was beginning to lose its charm. My affection for honeybees and the vast world of wisdom they bestowed upon me was enticing. It occurred to me that I was a worker bee, but longed to be the queen.

Saturday finally arrived, and I woke early with anticipation. Since I did not have the traditional protective beekeeper's jacket and veil yet, I had been advised to wear long sleeves and light colors. Bees tolerate light colors, which have the added advantage of keeping you cool while working under a hot sun. So, dressed in a long-sleeved white T-shirt, jeans, and my familiar black cowboy boots, I hopped into my Jeep and drove a few towns over to the apiary that was hosting the inspection.

The apiary was no more than twenty-five minutes away from my house and located somewhat off the beaten path, away from other homes. At eleven o'clock sharp I arrived at the end of a winding dirt road to find beekeepers, in full costume and with hive tools and

The Scientific Classification of a Honeybee

Kingdom: Animalia

Phylum: Arthropoda

Class: Insecta

Subclass: Pterygota

Infraclass: Neoptera

Superorder: Endopterygota

Order: Hymenoptera

Suborder: Apocrita

Family: Apidae

Subfamily: Apinae

Tribe: Apini

Genus: Apis

In 1758, Carolus Linnaeus gave the honeybee its scientific name, *Apis mellifera*. In Latin it literally means, "the honey-carrying bee." A more accurate name, *A. mellifica*, "the honey-making bee," was introduced in 1761. The Linnaeus name remained because of rules dictating scientific nomenclature, but *A. mellifica* is occasionally used in older bee-related literature.

smokers in hand, clustered to one side of an ivy-covered stone wall. This place was a pollinator's paradise, and crocuses, daffodils, and snowdrops bloomed in every direction as far as the eye could see. The gardens were exploding with blossoms. The beekeepers greeted me, and the host, named Billy, handed me a beekeeper's veil and hive tool: a piece of metal equipment that looks like a small crowbar. Billy told

HIVE TOOL

the group, "If you go to your apiary without your hive tool and smoker, go back and get them. There's no sense in even trying to open your hive without them." That was good old-fashioned beekeeping experience talking. He then led the group along the old stone wall and through an apple orchard—a favorite nectar-gathering spot, he said, for his apiary's honeybees. As we walked, he showed us the black locust and tulip trees that his bees liked in addition to the apple blossoms.

Billy pointed out an old cistern filled with water specifically for the bees to drink from. Curious how bees drink water, I wandered over for a peek. Floating on wayward leaves and twigs were the honeybees, sucking up cool water with their long tongues. A few bees had fallen into the cistern and were scrambling helplessly in the water, attempting to grab hold of a floating twig or leaf for refuge. Figuring I could help, I offered my hive tool. One by one, they each took hold of my tool with all six legs, and I was able to carry them over to a floating leaf for safety.

Honeybee colonies need water as much as they need nectar and pollen, and they drink several gallons of water a day. Water is used to dilute honey for brood food and cool down an overheated hive. It is recommended that beekeepers provide a water source for their colonies. A shallow tub filled with water and placed near the hive will keep the bees from visiting a neighbor's swimming pool or sucking water from clothes hanging out to dry on a laundry line. Placing a few strategic twigs on the water's surface will give the bees something to land on.

As Billy's honeybees sucked up the water in his cistern, I watched them closely and examined their little bodies.

Anatomy of a Honeybee

A honeybee's body is made up of three parts: the head, the midsection or thorax, and the abdomen. The bee is covered from head to legs with fine hair that easily magnetizes particles of pollen. Two antennae are

The Most Popular
Types of Honeybees

Honeybees are not native to the United States. Although bees have been in existence for more than thirty million years, they first arrived on this continent with colonists during the turn of the century.

German Black Honeybees (*Apis mellifera mellifera*) were most likely the first bees to arrive in the United States. Although they are thought to be aggressive and slow to build up colonies in the spring.

Italian Honeybees (*Apis mellifera ligustica*) became popular in the United States because they were very good honey producers and resistant to most bee diseases. Originating in southern Italy, they are now the most common of all honeybee species in the United States. They are golden to reddish brown in color. They tend to rob other hives while eating their own stored honey quickly.

Carniolan Honeybees (*Apis mellifera carnica*), from the Austrian Alps and the Danube Valley, are the second most popular bees among U.S. beekeepers. This breed of honeybee is known for swarming and is slow to build up combs. Maybe the gentlest of all honeybees, they are efficient in stocking up for the winter season.

Caucasian Honeybees (*Apis mellifera caucasica*), from the area near the Black and Caspian seas, are gentle yet they have a tendency to swarm and practically weld their hives shut with propolis.

Russian Honeybees originated in the Primorsky Krai region of south Russia. This breed is considered gentle and good nectar gatherers.

HONEYBEE ANATOMY

attached to a triangular-shaped head. The antennae are equipped with sensory hairs that allow the honeybee to touch, taste, and smell; these senses are important, since most of the honeybee's work is done inside a dimly lit beehive.

A honeybee has five eyes with a combined total of seven thousand hexagonal facets that detect movement, color, and light. Two large compound eyes are sensitive to movement and color, while three smaller ones, located above the compound eyes and called *ocelli*, are sensitive to light, but cannot see images. Could these hexagon-shaped lenses be responsible for the way honeybees construct their hexagonal shaped honeycombs? I've always wondered this, but have not yet discovered the answer.

For chewing and kneading beeswax into honeycomb, the honeybee has a *mandible*, or jaw. To suck up honey and water, the bee uses its tubelike tongue, called a *proboscis*.

Two sets of wings and six legs are joined to the thorax. A honeybee's wings are thin and clear and have veins running through them. They flap 180 times per second, and on average, a honeybee can fly 15 miles per hour. During flight, the wings hook together with a tiny hook, called a *hamuli,* on the back wings. All six of its legs are segmented and used for walking or grooming. The hindmost legs are covered in hairy baskets, which the bees use to carry pollen. A honeybee's abdomen houses its reproductive organs and digestive system, as well as its stingers and wax-making glands. Like most insects, the honeybee does not have bones, but instead has a hard, protective outer covering called an *exoskeleton.* Honeybees have blood but no veins in their bodies, only a single aorta that pumps blood from their four-chambered heart into the head. From there it is circulated around the abdomen freely. They also do not have lungs. Air sacs, or openings, allow oxygen to enter the honeybee's body. It is then pushed through a tracheal system of small tubes that carry the air to each cell.

· · ·

As WE CONTINUED OUR WALK THROUGH BILLY'S GARDEN, he pointed out that many farms in the state of Connecticut produce fresh fruit and berries and sell their produce exclusively to local farmers' markets and to bakeries that make delicious fruit pies. Honeybee pollination is crucial to the sustainability and profitability of these farms. Fruits, nuts, and berries start out as flowers. If honeybees do not pollinate the flowers, the flowers will never become the apples, pears, blueberries, blackberries, pumpkins, cranberries, peaches, hazelnuts, chestnuts, almonds, and squash found in the many baked goods we all love. If we enjoyed indulging in these home-baked goodies, Billy warned, then we'd be smart to continue keeping honeybees. During his talk, all I kept thinking was, what would I do without my mother's apple pie? By keeping honeybees, I would be actually helping Mom by saving the world's apples.

The group trailed through the gardens until we reached the apiary, which comprised six beehives in all. Billy told the group he

HONEYBEE

Mom's Bee-Pollinated Apple Pie with Honey

INGREDIENTS:

2 9-inch pie crusts
5 cups peeled, cored, and sliced Granny Smith
and Macintosh apples
1 cup Red Bee® wildflower honey
1 teaspoon cinnamon
1 teaspoon vanilla
2 tablespoons unsalted butter

Position rach in lower third of the oven and preheat oven to 400°F. Fill a pastry-lined 9-inch pie plate with sliced apples. Pour honey over the apples and sprinkle with cinnamon and vanilla. Dot with small pieces of the butter. Cover filling with second crust and slice vent holes in top crust. Bake for 15 minutes, then reduce heat to 350°F. Continue to bake until top crust is lightly browned and filling is bubbling, 40-45 minutes. Serve warm and drizzled with more honey.

had been keeping bees for four years and beekeeping was the most rewarding hobby he knew of. Today's hive visit would be a general springtime inspection to identify the inhabitants of the hive: the queen, the workers, and the drones. And we would learn to find the very tiny eggs inside the brood nest.

Using some quick-lighting wads of paper, called smoker fuel, Billy lit his bee smoker. As he did, he explained that if your smoker goes out, you do not want to waste time relighting it while your hive is wide open and exposed to the external elements. Smoker fuel keeps

A SMOKER

things continually smoldering without wasting a lot of prep time. The wads of paper were made up of compressed cotton fiber and could be purchased at bee supply shops.

Grateful for this handy tip, we the followed Billy over to the first hive, where we all listened intently as Billy talked about the hierarchy inside the beehive. To me, hive life, with its royal queen and her attendants, worker bees that do all the chores, and the lazy drones, sounded just like a fairy tale.

LIFE IN THE HIVE

A typical honeybee hive in the height of the summer will have a colony of approximately 80,000 bees, most of which are the female worker bees. Each hive has one queen, who is the mother of all the bees in her hive and the only sexually developed female. All the activities of the hive revolve around the queen and her egg-laying schedule. The queen is the largest bee inside the hive and can be identified by her long, streamlined abdomen. It is longer than the abdomens of the female worker bees and thinner than those of the male drones. Often the beekeeper marks the queen, using a special waterproof marker, with a

Queen

Worker

Drone

colored dot on her thorax. This dot serves two purposes. First, it makes locating the queen in a hive of 80,000 bees much easier, and second, the color of the dot designates the year in which the queen was born. The color system, used by beekeepers all over the world, is as follows:

Blue: for years ending in 0 or 5
White: for years ending in 1 or 6
Yellow: for years ending in 2 or 7
Red: for years ending in 3 or 8
Green: for years ending in 4 or 9

In order to mark a queen, beekeepers use a special tool called a *queen catcher*, a small, spring-action clip (that looks like a hair clip), to gently pick her up. She is then coaxed into a small, clear plastic queen-marking tube. The other side of the tube is closed off with a screen. A sponge-covered plunger is pushed through the open side of the tube and moves the queen toward the screen. When her back is against the screen, the beekeeper takes the special queen-marking pen and, through the screen, marks the back of her abdomen. The ink takes a few seconds to dry and then the queen is placed back into the hive.

THE QUEEN AND HONEYBEE REPRODUCTION

All female honeybees begin their lives the same way: as fertilized eggs. The eggs are laid in the *brood nest*. The brood nest is usually found on the inner-most frames of the hive and is made up of a collection of cells, each containing an egg, that form an oval-shaped *laying pattern*. A small, arc-shaped area of cells above the brood nest is filled with pollen. Above the pollen, another arc reaches into the top two corners of the frame and is filled with honey. Only one egg laid by an existing

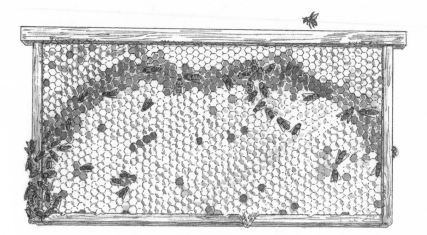

BROOD NEST IN THE TYPICAL OVAL PATTERN

queen inside a *queen cup* will develop into a queen bee. Queen cups are larger than the cells in which workers and drones are reared and hang vertically from the brood nest. The beekeeper can identify queen cups because they protrude, peanutlike from the brood nest.

For the first three days of their lives as larvae, all female honeybees are fed a diet of *royal jelly*, a protein secretion made from the hypopharyngeal glands located in the heads of mature workers. However, during the remainder of the larval stage, and for the rest of their lives, only potential queens are continually fed royal jelly, while female workers and male drones begin feeding on pollen and nectar. Royal jelly is what enables queen bees to develop into sexually mature females.

On the seventh day of the larval stage, the queen larva transforms into a *pupa*, and at that time the worker bees will

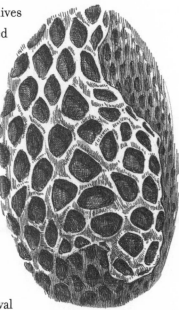

QUEEN CUP

close the queen cell with a beeswax cap. The young queen emerges approximately sixteen days later by chewing a circular cut in the wall of her cell until the wall swings open like a door. A handful of worker bees become her attendants and spend the rest of their short lives feeding and grooming her, cleaning up her waste, and following her around the hive. The queen truly lives the life of royalty. She does not gather nectar or pollen, make honey, or even take care of her young. These are the jobs of the worker bees. As a matter of fact, the queen rarely leaves the hive except to mate with the male drones or to swarm.

Within her first week, the newly emerged virgin queen leaves the hive for the very first time on what is called a *nuptial flight*. This dramatic event between the queen and drones from other hives happens in the air, well above the hive and the normal flight path of the worker bees. Drones gather in *drone congregation areas*. It is not known how these areas are decided upon, only that they are places away from the hive where the drones meet up and wait for a queen to fly by. The fastest drones are able to catch a queen and mate with her. To attract drones, the queen secretes a series of pheromones from her mandibular gland, called *queen substance*. The queen may mate with up to twenty drones. She makes only one mating flight in her lifetime, because that one flight gives her a lifetime supply of sperm. She stores the sperm in her *spermatheca*, her female reproductive organ that accepts and carries all the sperm. Upon returning to the hive, the queen begins laying eggs.

After she mates, the queen's glands mature and become functional, giving off pheromones that regulate all the undertakings of the beehive. Pheromones are chemical messages, essential to honey-

LARVAE FLOATING IN ROYAL JELLY INSIDE THE BROOD NEST

bee communication, that trigger certain kinds of activities in the hive. The queen's attendants distribute her pheromones to every bee in the hive by touching the workers around her with their antennae and by fanning their wings. These workers, in turn, touch the bees next to them and so on. The queen is the most important bee inside the hive, and each bee within the hive feels her influence. If a queen disappears or leaves the hive, it does not take long before her pheromones fade and the workers realize that their queen is missing.

During summer months, a queen lays an egg every twenty seconds, day and night, with a break every twenty minutes. She might lay as many as two thousand eggs per day—an amount that is approximately equal to her own weight. Healthy, productive queens lay an egg in almost every cell in the brood nest and leave just a few cells empty in between. This is considered a normal brood pattern. Eggs can be difficult for beginning beekeepers to spot; they look like small pieces of rice standing straight upright, and there should be only one inside each cell. In the larval stage the eggs look like shiny white worms curled into the shape of a "C" and floating in royal jelly. Once the cells are capped, the larvae spin a silky cocoon and straighten out again, becoming pupae. Cappings are the convex beeswax seals made by worker bees to cover or cap the cells. Once the egg cells are capped, the larvae actually begin to look like bees.

Not all of the eggs get fertilized. The unfertilized eggs become males or drones, while the fertilized eggs become female worker bees. If a queen never mates or simply never becomes fertile, all her eggs will become drones, and she should be replaced by the beekeeper. A queen bee can live up to five years, but her egg production will begin to drop off after two or three years. For this reason, most beekeepers must eventually replace, or *requeen*, their hives. Requeening each year ensures that a hive has a young, healthy queen; a strong queen, in turn, means a strong colony that generates maximum honey production.

The queen bee has a stinger, but it is not barbed as a worker's. She is capable of stinging repeatedly, but reserves her sting for desperate moments of combat with other queens. She does not use it to defend the hive.

When we opened up Billy's hives, we observed several honeybees hanging around, just flapping their wings. This behavior is known as *fanning*, and in addition to distributing the queen's pheromones, it circulates air in order to maintain the proper temperature inside the hive. During brood rearing, the temperature of the brood nest is maintained at 90 to 95°F (32 to 35°C), even if the outside temperature is above 100°F or below zero. Since worker bees metabolize honey to generate the heat needed to warm the hive, honey must be present in the hive at all times.

The presence of a queen is essential to life in the hive. A colony will raise a new queen for a few different reasons: if the existing queen is removed or accidentally killed by beekeeper's error, or lost on her mating flight, or if an older queen is failing and her production of queen substance and egg laying declines.

Queens that are killed, removed, or lost will be replace by the colony with an emergency queen. The colony will select a young fertilized egg or young worker larva that is not more than three days old. This specially selected egg or larva will be in a worker-sized cell that will be enlarged to a queen cup to accommodate the queen's larger body.

A failing queen will be replaced by an existing queen in a process called *supersedure*. The queen will lay a fertilized egg in a queen cup. When the new queen emerges, the old queen will be *balled* by the workers; a process in which the workers cluster around the queen until she dies from overheating.

Queens reared by supersedure are usually stronger and better cared for than emergency queens since they are not created in a panic and will receive larger quantities of food (royal jelly) during development.

WORKERS

Worker bees are all female, and they comprise the largest group within the hive. They are sexually undeveloped and do not lay eggs under normal hive conditions. However, worker bees do everything else, from feeding and grooming the queen to gathering pollen and nectar and raising brood. They also guard and clean the hive and, most important, make the honey.

A worker bee begins her life as a fertilized egg, which hatches into a larva. As the larva grows it spins a cocoon, and through the process of metamorphosis turns itself into a pupa. Throughout this maturation process, the larva is completely dependent on the care of adult worker bees.

Workers are fully developed and emerge from their cells in twenty days, which is four days less than it takes for a drone to fully develop.

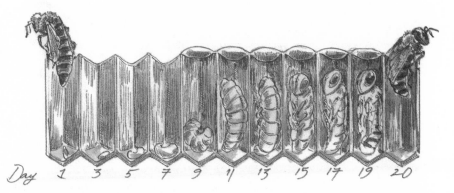

BIRTH CYCLE OF A WORKER BEE

The average worker bee lives approximately four to six weeks in the spring or summer and up to six months during the winter. Each worker specializes in one specific task at each stage of her brief life. The first few days after she emerges from her cell she serves the hive as a house bee and a cell cleaner. She inspects, cleans, and polishes the cells in which the queen lays her eggs. She then begins her duty of feeding older larvae with pollen and nectar brought into the hive by more mature worker bees, as well as with a substance called *bee bread*, which is a mixture of fermented pollen and honey. At ten to sixteen days, her glands develop, and she begins to secrete and feed royal jelly to the younger larvae. As her wax glands develop in her abdomen, she also secretes beeswax and uses her mandible to form it into comb, a series of beeswax cells that make up the brood nest and in which pollen and honey will be stored. This beeswax is also used to cap the cells that are filled with larvae or honey. At this stage a worker bee also helps to repair damaged comb. In between these activities she takes time out to flap her wings to circulate air and provide ventilation inside the hive and to help removed moisture from the pollen, brought in by foraging bees, in order to create honey.

At about twenty days old, the worker bee will become a guard bee, stationed at the entrance of the hive. Guards admit returning bees only if the bees are part of their hive. All bees must carry their own queen's pheromone to gain entry to the hive. Guard bees also reject old or diseased bees and drive out the drones in the fall. Guard duty is a worker's last job inside the hive before she becomes a forager bee. By this time, the worker bee's sting glands are fully developed, and it is her duty to defend the hive. A honeybee dies after inflicting one sting.

A worker bee does not begin foraging for pollen and nectar until she is three weeks old. At this point she will leave the hive to collect nectar to make honey, pollen to feed the hive, water to drink, and *propolis*, a resinous mixture used to seal up cracks and other unwanted spaces in the hive. On average, a single worker bee makes only $\frac{1}{12}$ teaspoon of honey in her lifetime.

With all these hive duties, a worker bee truly works herself to death, both by wearing out her wings and through sheer exhaustion. It is not uncommon for adult bees to die inside the hive. When they do, it is the job of the undertaker bees to remove them.

DRONES

Drones are the male bees, and they make up only 10 to 15 percent of the total honeybee colony. Drones are reared in the same fashion as the queen and worker bees, except they develop from unfertilized eggs. The drone cells are found at the lower part of the brood nest, are larger than worker cells to accommodate the wider male body, and they have raised wax caps that look like a bullet. Drone eggs hatch in three days and emerge from their larval stage in five to six days. The drones' pupal stage is five days longer than the workers'. Chubby and squat in stature, adult drones emerge from their cells fully developed in twenty-four days and have very few duties inside the hive. They do not gather pollen or nectar and do not make honey. They do not clean the hive or take care of the young. They cannot feed or groom themselves. Because they do not have stingers, drones cannot even defend themselves or the hive. However, drones are important to the hierarchy of the hive.

A drone's sole responsibility is to mate with a virgin queen from another hive, and mating is one of the only reasons drones leave the hive. Drones wait at designated drone congregation areas—something like a drone hangout—for a virgin queen to fly by to mate with. The

Birth Cycle of Honeybees

	Egg Hatch	Larva	Pupa	Emerge
QUEEN	3 Days	5.5 Days	7 Days	16 Days
WORKER	3 Days	6 Days	11.5 Days	20 Days
DRONE	3 Days	6.5 Days	14.5 Days	24 Days

BEE LINING
How Ancient Man Found Honey

Before man domesticated bees, *bee lining* was an attempt at locating a feral bee nest to hive or just some wild honey as a sweet treat by following foragers back to their home. Today, this ancient practice, also called bee hunting, is still a fun activity for a nice day with fellow beekeepers. I joined one hunt while attending a beekeeping gathering.

You begin by locating a few honeybees at a flower patch or watering hole. Then the goal is to try to lure them with some sugar syrup into a box. Once they are in the box, you mark each bee with a color dot on her back, similar to the way you would mark the queen of a hive. Honeybees will naturally return to their hive once they have filled up on sugar syrup and recruit other foragers to the source. After the bees have been marked and have filled up on sugar, you can release each bee at different areas and watch as they return to their hive. The direction in which the bees fly will determine the general location of their wild nest. If you successfully find their nest, you can be sure by spotting the marked bees. If the sun is at an angle, such as in the early morning or late afternoon, it will reflect on to the departing bees, making it easier to follow them while they're in flight and thus determine the direction of their travel.

Ancient honey hunters may have had to climb dangerous trees or mountains to actually get the honey, and without proper protection, they must have endured a few stings. Bee lining can be challenging the first few times, and it is easiest to do when the temperature is fairly warm and the bees are out foraging in full force.

fastest and strongest drones successfully mate in midair, copulating with the queen from behind in a brief and rather violent encounter. The act of mating concludes with the drone's genitalia being ripped from his abdomen as he falls to his death. (Sounds like a one-act opera, doesn't it?)

Most of the drones are not allowed to spend the winter in the hive, because they would eat all the stored honey. So, sadly, drones are forced out of the hive in the autumn and generally die of starvation or cold. It is not uncommon for beekeepers to see the drone corpses at the hive entrance in late autumn.

• • •

ON MY WAY BACK HOME FROM BILLY'S APIARY, I spotted a single honeybee buzzing around the back of my Jeep. Somehow she managed to find her way in and was now on a one-way trip across town. I felt terrible knowing she would not make it back home tonight, or maybe never make it home at all. What would she do? Where would she sleep? Promptly, I rolled down the back windows, hoping she would find her way outside and back to her hive. She was sucked out by the blast of wind, and I will never know if she made it back safely. It was just the kind of thing that happens all the time, but I'd never thought about it from the bee's perspective until that day. Since honeybees are social creatures, I'd learned from Billy, they rely on the colony as a whole to survive. They return home each night, but tonight where would this bee go?

Later, I remembered to ask Mr. B about that incident. Mr. B assured me if there were a hive in a reasonable distance, her chance of being accepted would increase if she entered with an offering of pollen or nectar. That thought was comforting. I could only hope she found a hive that welcomed her that night.

CHAPTER 4

Foundations in Beekeeping

A few days after the hive inspection, Mr. B called to ask me if I was ready to set up my first hive. If so, I would have to order my colony of bees in the next few days. It was already late April and the time for me to make my decision was running out; bee-breeding farms would not ship bees in the warmer summer days because of stress on the bees and heat-related losses. At that moment I took the plunge and committed to keeping honeybees. I was ready as I ever would be and knew Mr. B would be there every step of the way. I had officially become a beekeeper.

Now that I was a beekeeper I would need a beehive. There are many suppliers of beekeeping equipment in the country, and one can order equipment from them online or through one of their many catalogs. I picked up a few free catalogs at my beekeepers meetings to get a taste of what was available.

Turns out there are many styles of hives to choose from and an abundance of accessories. I opted for a beekeeping starter kit, which included a standard Langstroth-style wooden beehive and all the basic beekeeper's tools. I could always add more beekeeping goodies to my collection later on as necessary.

The Langstroth Hive

The Langstroth hive is the style of beehive that most beekeepers in the United States use today. Prior to the introduction of the Langtroth hive, bees were kept in hollow wooden logs, called gums; in clay pots; or in woven bee skeps. With any of these primitive beehives, it was difficult for beekeepers to manage their bees, find their queen, and prevent diseases. Honey harvest was the most challenging time, and when it came time to collect the honey from inside the hive, there was little choice but to set these primitive beehives on fire, killing the queen and destroying the entire colony, in order to remove the bees from the honey. Then, to enjoy the honey, people had to squeeze it from the comb by hand or just chew the comb itself.

The Reverend Lorenzo Langstroth is known affectionately among beekeepers around the world as the father of beekeeping. In 1852,

WOVEN BEE SKEP

Langstroth's innovative hive design revolutionized beekeeping with his use of removable frames that hang inside the hive box. They were loosely based upon a system invented by another beekeeper, a Ukrainian named Petr Prokopovich. These wooden frames on which honeybees build their comb are lined up in neat rows and are easily accessed by the beekeeper, who only has to open the top of the box and lift out a frame in order to inspect and manage the hive and collect the honey. In 1789, before Langstroth's innovative design, Swiss inventor François Huber designed a hive that opened and closed like a book, with frames bound in like pages. It was called a Leaf Hive. But it was Langstroth's discovery of *bee space* that brought beekeeping into the future.

Langstroth wrote about this discovery in his 1853 book titled *The Hive and the Honeybee*. He discovered that if there was more than ³⁄₈ inch of space anywhere within a hive, the bees would build comb in that gap, which he called bee space; if there was less than a ³⁄₈-inch space, the bees would fill the gap with sticky propolis, making the beekeeper's job of removing frames much more difficult. Langstroth's hive design revolutionized beekeeping because it had movable frames, spaced not less than ³⁄₈ inch apart, which addressed the concept of bee space. In 1859, a German named Johannes Mehring produced the first premade beeswax foundations for Langstroth's frames. Embossed with a honeycomb pattern, these foundations gave honeybees a head start in building honeycomb. Then, in 1866, Austrian Franz von Hruschka, after observing milkmaids swinging buckets of milk in circles, invented the first device to extract honey from the comb using centrifugal force.

• • •

MY FIRST BEEHIVE ARRIVED THREE DAYS AFTER I ORDERED IT. When the delivery truck pulled into my driveway, I ran out to greet it like a kid runs to an ice cream truck. The hive came packed in two large, heavy boxes with charming illustrations of honeybees printed on the sides.

I carried each box to my back patio. Using a box cutter, I meticulously sliced through the top of the first box where the seams met, and then unfolded each flap to expose the aroma of fresh pinewood and beeswax.

The precut wood was packed snugly with the galvanized nails. There were also twenty sheets of beeswax foundation, wrapped between cardboard for safety. The second box held the beekeeper's tools of the trade. Wrapped in protective paper was a beekeeper's hat. The hat was tan and resembled a woven safari helmet with a black mesh veil that protects the face and neck from curious honeybees, making visiting the honeybees seem like a religious experience. Also included in the second box was a white protective beekeeper's jacket, which had a zipper around the neck to attach it to the veil, and the indispensable hive tool. The hive tool is primarily used to pry open hive lids that have become stuck shut with beeswax or propolis.

I also found a pair of beekeeper's leather gloves that were long enough to reach my elbows. The last item I pulled out of the box was a smoker, which looks like large stainless steel can with bellows. Beekeepers light a small fire inside these canisters and squeeze the bellows to draw air into the flame. A few pieces of newspaper, dried leaves, and twigs can quickly transform a smoker into a fireplace, but beekeepers agree that keeping your smoker lit during an entire hive visit is possibly the most difficult part of beekeeping.

Assembling the Hive

A Langstroth beehive looks a lot like a wooden file cabinet. But its rectangular boxes, stacked neatly on top of one another, are not file drawers. They are called *deeps* or *hive bodies* and they hold the wood frames on which the colony of honeybees live, build their honeycomb, and raise brood. Each deep has four sides, but no top or bottom, since they sit on top of one another. Grooves on the outside of the deeps act as handles for lifting. My deeps were the standard 9⅝-inch size. I later learned some beekeepers opt for *medium-depth* hive bodies that are 6⅝ inches deep. Since these medium hive bodies hold frames that are shorter, they weigh less and make lifting or maneuvering them easier on the beekeeper's back.

Putting together my first hive might have been quicker if I'd been able to watch someone else do it first or if I'd had some experienced help. But I was surprised at how easy mine was to assemble. I was able

TYPICAL LANGSTROTH HIVE

to finish it using only a hammer and a little elbow grease, especially since the precut main parts simply locked together, tongue-and-groove style. I also used a little bit of wood glue to reinforce all the joints, which I knew would help the structure last longer.

Each of my deeps held ten removable rectangular frames, which are referred to as *deep wooden frames*. The wood for the frames, like the wood for the deeps, came precut and the pieces locked together easily. The last step was to insert a beeswax foundation into each, like a picture inside a frame. Getting those sheets of beeswax foundation to neatly slip into the frames was tricky. A slit at the top and bottom of the inside of each frame held the sheet in place, and small clips called *foundation pins* held the sheets to the sides of the frame. Once in place, the sheet was secured at the top by a thin piece of wood nailed across the top of the frame. (I found that a brad driver came in handy here for the tiny nails.) As any new beekeeper will tell you, placing the wax foundation sheets neatly inside the frames without damaging the wax is a challenge until you've done it a few times. Fortunately, if you bend or crack the wax foundation by accident, the bees will repair it once it is inside the hive. They will fill in a beekeeper's clumsy holes with their own beeswax to create a perfect honeycomb pattern. Bless these creatures!

The foundation sheets were embossed with a honeycomb pattern, and my honeybees would eventually add beeswax to the foundation to make their cells or comb. It was there that my queen would lay her eggs and the brood would be raised. A thin metal wire running through the foundation would support the weight of the honeycomb as the bees built it up.

My completed hive body had ten frames in each deep, for a total of twenty frames. Along the top of the deeps were edges called *rabbets*, from which the frames hung in a neat line. I nailed thin, L-shaped pieces of metal along both sides for protection. With the frames sitting on two of these metal guards, it would be easy for me to slide the frames in and out of the hives during harvesting or inspection without crushing the bees.

My beehive had an *outer cover* to protect the top of the hive and shield the bees from nasty weather. Sometimes referred to as a *telescoping cover*, this wooden board was covered by a thin sheet of

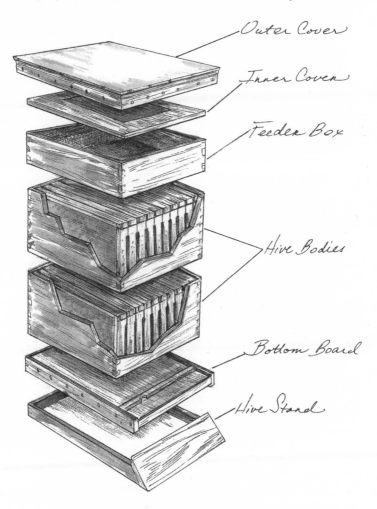

Outer Cover

Inner Cover

Feeder Box

Hive Bodies

Bottom Board

Hive Stand

FOUNDATIONS IN BEEKEEPING

galvanized metal that hung over the sides of the hive body. Underneath this outer cover was a flat piece of wood with a framed edge, called, cleverly, the *inner cover*. It would serve as extra insulation for the bees and prevent them from building honeycomb on the outer cover. On the front side of the inner cover was a small open notch and, on the top, an oval hole. Both openings acted as vents, so moisture could escape the hive.

At the base of the bottom deep was a platform called a *bottom board*. This solid piece of wood had three edges for the bottom deep to sit on. Serving as the base of the hive, the bottom board, which stuck out a few inches from the rest of the hive, would also be an entrance or landing pad for the bees. (Later, I would learn about screened-bottom boards that could be used for Varroa mite control.) The *entrance reducer* was a long, thin piece of wood with different-sized holes, and it sat at the entrance on the bottom board. It would control the size of the entrance, where the bees would go in and out, and help regulate ventilation inside the hive during various seasons. The *feeder box* sat between the deeps and the inner cover. My feeder box was a wooden tray that held the sugar syrup for feeding my bees. At one end there was a screened area for bees to crawl up to from inside the hive, where they could safely drink the sugar syup without drowning in the tray.

The entire setup rested on top of a *hive stand*, which was a bit more complicated to assemble because of the angles of the wood I needed to hammer together. The whole process of assembling my first hive and frames took almost five hours. Experienced beekeepers can do it in two.

Once the hive was completely assembled, I was ready to prime and paint it. I chose to paint my hive a carnelian-inspired red to harmonize with my red cottages. Only the wood on the outside of the hive needed painting. I used two coats of water-based acrylic and a clear, protective shellac for added protection. I left the hive entrance unpainted, as bees like it au natural.

Staging the Apiary

The next step was to choose a place in my yard for my new beehive. Location, location, location is important to the success of a hive. Mr. B advised me to situate the back of the hive facing north, so the colder winter winds would not blow directly into the entrance. A fence, hill, or row of trees could aid as a blockade. He also told me to find a place where the hive would get early morning sunlight, because then the hive would warm up more quickly, which would prompt the bees to get an early start foraging, and to avoid damp places where water accumulates, but to create a source of water for the bees to drink.

The entrance of a beehive is like a runway at an airport, and honeybees would be constantly flying in and out. I didn't want visitors, children, or pets passing directly in front of their flight path, but it was also important to me to place my hive somewhere visible from a house window or two. It would be nice, I thought, to glance over and observe my bees while I was working or cooking. I settled on a spot just outside my kitchen window and toward the back of my yard. From where my honey shrine was situated, I could eat my honey and watch the bees that made it.

CHAPTER 5

My Queen and Her Subjects Arrive

Numerous bee farms around the country sell packages of honeybees, and each package contains a colony of 18,000 to 20,000 bees, including one queen. Some bee clubs will bundle orders together, so all their members can pick up their colonies at one location. Mr. B suggested I order my bees from an old friend of his in Georgia.

My first colony of Italian honeybees was delivered through the U.S. mail service on a Saturday in mid-May. The post office phoned that morning to let me know the bees had arrived and to beg me to *please* pick them up as soon as possible. When I arrived at the post office and hour or so later, I was asked to go around to the back door to retrieve my package so as not to alarm the other customers. When I told the clerk at the back of the building that I was there to pick up my bees, she jumped up and darted into the back room, yelling, "Loouie! They're here to pick up their beeees!" Louie, her fellow employee, emerged from a back room wearing a pair of barbecue mitts and carrying my box of bees at arm's length. He looked incredibly relieved to be handing them over to me. I accepted the wooden box, which looked like large shoe box with mesh screens on two sides. The bees could definitely be seen and heard through the screened sides. Day-Glo stickers were plastered all over the box, declaring "LIVE BEES, CAREFUL!" (I later peeled off one of these stickers and saved it as a memento.)

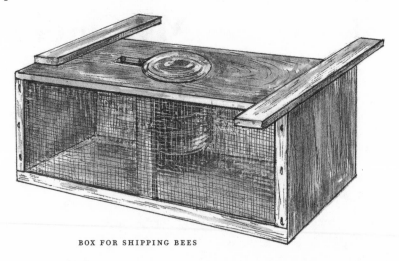

BOX FOR SHIPPING BEES

I carefully placed the box of bees in the hatch of my Jeep and sprayed them through the screen with a sugar-water solution I'd mixed at home. The sugar, besides feeding the bees, also helps to calm them by filling their bellies. While licking the sugar off their bodies, they become fat and happy—a form of honeybee intoxication. Inside the wooden box I could see that the bees had formed a tight cluster that appeared to be hanging from the top of the box. They let out a gentle and steady hum that sounded content. I fastened the box of bees with a bungee cord to the floor of the Jeep and closed the hatch door. Once they were all tucked in, we were on our way. I drove home, taking care to not bounce the bees around and upset them, yet all the while I felt provocative, knowing I was driving around with live honeybees in my truck.

We pulled into the driveway, all 20,001 of us. Before removing the box from the truck, I gave the bees one more squirt of sugar water. They still seemed calm as I carried them into my cool basement. They would be safe away from direct sunlight or cold breezes that might give them a chill. I prepared another batch of sugar-water syrup using one part white granulated sugar to two parts water. To mix it up, I just had to boil the water, turn off the flame, and then add the sugar. I stirred it well until the sugar dissolved completely, and then let it cool before giving another serving to my bees. Tomorrow, Mr. B was coming over to meet my girls, as he called them, and to show me how to properly hive them. I had read over and over about how to properly hive a package of bees, carefully reviewing the steps and the photos. I'd even watched a video loaned to me by the bee club. But I was grateful that there would be an expert on hand for my first time.

Other Ways to Acquire Bees

Many new beekeepers are steered toward buying packaged bees, since that's the easiest and most common way to establish a new hive. However, honeybees can be obtained in other ways besides ordering them through the mail. Some beekeepers sell frames containing worker bees, a mated queen, and existing new brood. You simply put these frames inside your hive, and the bees do the rest. This frame is called a *nucleus hive,* or simply "a nuc box," and gives the colony a jumpstart, because you don't have to wait for the queen to mate and lay eggs. Another way to acquire bees is to capture a swarm of feral bees, but wild bee colonies are rare in the United States. Finally, some beekeepers with established hives simply split an existing colony.

THE FOLLOWING AFTERNOON, I CHECKED MY BEES for what seemed like the fiftieth time and carried them up from the basement to the patio to enjoy the warm sunny day while they waited for their move to the hive. It was Mother's Day. What day could be a more perfect day to meet the queen and hive my honeybees? It turned out that in the Northeast "bee-hiving day," as we beekeepers call it, always seems to fall on Mother's Day.

Mr. B would be along any minute. I reviewed my notes and prepared my smoker with some newspaper and twigs. My hive tool, bee veil, and beekeeper's jacket were ready to go. I wore blue jeans tucked into work boots, so bees could not fly up my pants legs. I showered, washed my hair with fragrance-free shampoo, and brushed my teeth, because I had read that honeybees react unfavorably to body odor and perfumes, as well as the smell of leather or wool. Apparently bees sense these odors as similar to those of a wild predator, and may become agitated if the beekeeper smells like a threat to them. This is why beekeepers often do not visit their beehives wearing leather shoes, boots, or watchbands.

At exactly eleven thirty, Mr. B's old pickup truck ambled into my gravel driveway and stopped abruptly up against the old stone wall. The driver's side door opened, and Mr. B climbed out empty-handed. "Hmm, no bee veil," I thought. He smiled and walked directly toward the hive, motioning for me to bring my box of bees and follow him. He was clearly not going to leave me any time to be nervous.

When we arrived at the hive, he briefed me on the sequence of events for getting the bees into the hive and on the special handling of the queen. Then he handed me two small hive-frame nails and asked me to bend them to ninety degrees with a rock. I had forgotten what these nails were for, but I obliged. Next I removed the outer and inner hive covers, and the top deep, as well as five frames from the bottom deep, in order to create space for three pounds of live bees to be poured into the hive. Three pounds! The idea was thrilling and dreadful at the same time. Then, with the hive ready, my moment had come. I held the box of bees by its corners and banged it on the ground a few times, so the bees would fall to the bottom of the box. Oh, they did not like that one bit! Next, using my hive tool, I pried the small wooden plate off the top of the box and removed the tin can full of sugar water that had been placed inside for feeding. Then I replaced the wooden plate on top so the bees could not escape. Once the can was out of the way, I slowly slid the wooden plate to the side again and removed the queen cage. Once again I slid the wooden door back in place.

Hiving the Queen and Her Colony

The queen cage is a small box, not much larger than a pack of chewing gum, with a screen on one side and a sugar cork sealing it closed. Inside the box, the queen travels safely along with five or six female attendants who groom and feed her. This cage needed to be removed from the main package and placed in the hive before the rest of the colony. Now here was where those two bent nails come into play. Mr. B instructed me to stick them into the top of the queen cage so that it could hang between two frames inside the hive. With another nail I poked a hole into the sugar cork. Over a period of about a week, the worker and attendant bees would eat away the sugar cork until the queen was released from her cage. Then she would be free to meet her family and spread her pheromones throughout the hive. I hung the queen cage with care, with the sugar cork facing up, in the space where normally the fifth hive frame would fit.

The easy part was over. Now it was the time to open the bee package and pour the bees into the hive. I gave the bees one more squirt of sugar water for good luck, and then I grabbed the bee box with both hands and once again pounded it on the ground. I took a deep breath and moved aside the wooden plate for the last time, exposing a circular hole the size of the sugar can that was removed earlier. Immediately, bees began to crawl up and out of the hole. I had to work fast. I stood, feet planted to the

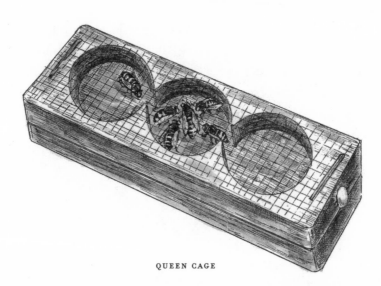

QUEEN CAGE

side of the open hive, turned the box upside down, and begin shaking and pouring the bees in as quickly as possible. That box was heavy and boiling over with thousands of live, seemingly unhappy honeybees.

I wish I could say that my first hiving experience was uneventful, because, in truth, most hiving experiences are. But as it turned out, my bees were not in the mood to be poured that day. Instead of funneling into their new home in an orderly fashion, as is usually the case, my bees became defensive and agitated. These were not like the docile Italian honeybees I met last month at Mr. B's apiary. These bees crawled all over my veil and began to sting me through my blue jeans. Unfortunately, once a bee is provoked to sting, she releases an alarm pheromone warning other workers that danger is afoot. This is the signal for the other bees to defend themselves, their honey, and their hive, and more stinging ensues. A few curious bees had already found their way under my veil and down my bee jacket.

I panicked, which was exactly the wrong thing to do.

I screamed and ran away from the hive, removing my veil and even my bee jacket. I could hear the bees buzzing in my hair as if trapped in a spider's web. Mr. B shouted after me to remain calm, and that the more movements and swatting I did, the angrier they would become.

But I was already frantic. I screamed, "Get them out of my hair. Help me!"

Mr. B took his hive tool and began flicking the bees out of my hair. Finally, not without several more stings, they were out. Mr. B. showed me how to remove the stingers from under my skin. Rather than pulling them out with a tweezers, it's quicker to scrape them out with your fingernail. The quicker the stinger is removed, the less venom will enter at the point of contact.

I needed a moment to compose myself, so I stepped away from the hive and shook out my veil one more time. When I was ready, I put my veil back on and carefully closed up the beehive. I put the feeder box on top of the hive and filled it with sugar water. At Mr. B's suggestion, we left the partly empty package of bees in front of the hive entrance, so the last few stubborn creatures would follow their queen's pheromones and thus find their way into the hive before dark. Last of all, I placed the entrance reducer on the bottom board of the hive with the smallest opening. I gave a sigh of relief and decided that I was finished with my beekeeping for the day.

At last, I found the courage to look over at Mr. B.

"What happened?" I asked.

Mr. B. asked me if I was wearing any perfumes, if I used a particularly fragrant shampoo that morning, and what I ate for breakfast. I responded in defense that I did clean up that morning, but was careful not to use anything fragrant. Breakfast was the usual cup of espresso with Mr. B's own honey, and cereal with bananas and honey on top.

"Oh, no. Not bananas," Mr. B groaned.

I blushed like a child being scolded.

Studies have shown, Mr. B explained, that a bee's alarm pheromones smell similar to bananas. It's possible that my bees smelled the

Stinging Behavior

Honeybees simply do not fly about looking for someone to sting. They are too busy collectiing pollen and guzzling nectar. Stinging is an act of defense. Beekeepers know this and acknowledge stings as a part of keeping bees. In time, most beekeepers become indifferent and immune to stinging.

Stings by wasp or yellow jackets are more common than honeybee stings. Honeybees will sting only if they feel their hive is under attack. The guard bees are the first to take notice of an intruder at the hive. Their duty is to protect the hive from danger and to alert the other bees. The moment a honeybee stings, it is the end of her short life. Although wasps and other bees can sting more than once and survive, when the honeybees stings, her barbed stinger, along with other vital body parts, is ripped out of her abdomen, and she dies. The longer the stinger remains in your skin, the more venom is pumped from it. Quick removal cuts down on the amount of venom you receive, which, in turn, decreases the pain or swelling.

Due to the alarm pheromones that are released after the sting, it's likely that other bees will come after you. It's important to stay as calm as possible and to not swat at the bees. Honeybees do not detect slow-moving objects, so standing still or slowing stepping away from the hive is the best way to avoid further stings. And you can always use your smoker to apply smoke to the sting, thus masking the alarm pheromone and confusing the bees.

Very few people are highly allergic to bee stings and show severe signs of *anaphylactic shock* when stung. Symptoms include shortness of breath, abdominal pain, vomiting, dizziness, and diarrhea. If you do get stung and are allergic, scrape the stinger out of your skin as quickly as possible, using your finger or a credit card, and seek medical assistance. Beekeepers are wise to have an EpiPen or an injection of epinephrine handy for any emergencies and for anyone showing these symptoms after a bee sting of any type. Apiary visitors who are allergic to bee stings should simply keep their distance from the bees and keep their sting kit, including any prescriptions from their doctor, with them. Being allergic to bee stings does not necessarily mean that you cannot or should not keep bees. I have met beekeepers who are highly allergic to bee stings; they completely protect their skin with a full-body beekeeping suit, gloves, and veil at all times.

banana I'd eaten earlier—perhaps there'd been some residue on my hands or clothes—and interpreted it as alarm pheromones. Who could blame them for electing to defend themselves?

My official sting count for the day was six. The stings swelled up like little red balloons and were seriously itchy—worse than any case of poison ivy I've ever had. But they were only painful for the first twenty-four hours and were gone in a few days, and they didn't deter me from working with honeybees. Since my first eventful hiving, I have hived dozens of packages of bees and never have I experienced a hive as hotheaded as those Italians. For some reason, they were agitated, and I got the brunt of it. I've chalked it up to beginner's luck.

After hiving my honeybees (or something of the sort), we let nature take its course. In one week I would return to see if the sugar cork was eaten and the queen was released from her cage. I was looking forward to my first inspection of my very own honeybees.

CHAPTER 6

First Observations as a New Beekeeper

The next day I was delighted to walk over to my bee yard and find that the last few of my bees had made it out of the bee box and into the hive. The weather was perfect—close to seventy degrees—and the morning sunlight cast a dappled pattern upon my red beehive. A few bees gathered at the entrance while dozens of others hovered around it. Many others flew in and out of the hive. A low, soft hum floated on the air. The incredibly rapid beat of the bees' wings and the vibration of their thoraxes create this familiar sound we call buzzing. For musicians, it is a steady, single note in the key of C-sharp. Signs of life, I thought. A sense of peace and comfort that I cannot quite explain in words came over me. Though there were hundreds of honeybees buzzing around my head and an abundance of activity at the hive entrance, there also was a distinct sense of order and serenity about it all. Anyone who has kept honeybees knows this sense of harmony. I stood in front of the hive, quietly observing my bees. After a few minutes, I needed a chair. The activity was all so hypnotic, I could not pull myself away.

The short flights my bees were taking around the hive are called *orientation flights*. It appeared that the workers were simply hovering around the hive, but as the name implies, these flights allow the bees to get acquainted with the location of their new home. Worker bees help their sisters return home by distributing another pheromone from their *nasonov* gland located on the tip of their abdomen. Standing on their back legs and lifting their abdomens, the workers flap their wings to release the scent. I noticed the bees at the entrance of the hive engaging in this interesting behavior.

The bees I saw in flight were definitely the worker bees, more slender and more graceful than the drones, who seemed to just be walking aimlessly around the hive entrance. Occasionally I'd observe a worker walking right on top of a drone or pushing him off the landing strip as if to say, "Out of my way! I have things to do!"

Each day for the first week after I'd hived my bees, I visited them and observed every one of their movements. I kept my notebook nearby to write down what I saw and draw a few sketches of my bees. They didn't seem to mind my presence at all. I noticed that on rainy days the bees stayed inside, but when the sun shined, they were animated and industrious. Before the week was out, I observed that the worker bees had begun carrying in brightly colored granules of pollen balls on both of their hind legs. Foraging for nectar and pollen meant the hive was full of activity and young bees were being raised. Some of the granules were positively huge relative to the size of the bee, and others were tiny. Yet, despite the granules' size, each bee seemed to carry them effortlessly into the hive. The workers entered the hive without objection from the guard bees who stood on the lookout at the entrance.

My workers seemed to have such a purposeful life, contentedly visiting the crocuses and daffodils, interrupted only by the occasional *cleansing flight*. Yes, honeybees leave the hive to find a restroom. They will never soil their pristine environment, thus ensuring a perfectly immaculate site for making honey. This fact makes one love these creatures more than ever. My honeybees passionately went about their duties, as if nothing else in the world mattered. If only I could be so content and focused on my own work.

First Peeks Inside the Hive

Exactly one week after I hived my bees, it was time to open the hive to check on my queen. It takes approximately three to five days for the queen to be released from her royal cage. The queen's attendants, who traveled with her inside her cage, and the workers inside the hive had been happily dining on the sugar cork that acted as a temporary door between the queen and her subjects. Once the candy cork was completely eaten, the queen would be released, and she would introduce herself to her subjects through queen substance. This pheromone—a chemical e-mail, if you will—is passed along from one bee to another through their antennae until it reaches the last bee. This is how the entire hive knows the queen is present. Once the queen was released into the hive, she would mate and begin laying eggs, and all the activities of a working beehive would officially begin. I was hoping that the queen inside my hive had been released during the first week, and I suited up with my bee veil, hive tool ready, to see how my bees were doing.

The best time to open up and inspect inside a hive is between ten in the morning and four in the afternoon. This is when most of the workers are out foraging. There will, therefore, be fewer bees inside the hive, which makes the job of manipulating the frames much easier.

First, I attempted to light my nice, new shiny smoker. I stuffed it with newspaper and a few twigs, struck a match away from the breeze, dropped the match in, and watched it flame out. I lit a second match, set fire to some newspaper, and then dropped it in. Using my hive tool, I moved the paper around in the canister and added a little air by squeezing the bellows. A miniature blaze broke out, but then quickly self-extinguished. Sigh. Lighting a smoker was not nearly as easy as it sounded. I needed a plan and some luck.

Again I tried the technique of lighting the paper first and then dropping it into the canister. This time I slowly added dried grass, more twigs, and leaves. After a five-minute battle, this method actually started to work, but it took some time to get those twigs to actually catch fire. I kept wondering, "How does a single match burn an entire house down when I cannot even light a humble fire in my smoker?" There unequivocally was an art to keeping your smoker lit, and it would be years before I mastered it. But eventually that day I established a smoldering flame.

Donning my veil, I was ready to open my hive. Standing to the side of my hive, with the sun at my back, I blew some smoke at the entrance. Then, with one hand, I lifted the front outer cover while puffing away with the smoker with the other. I removed the outer cover and leaned it along the side of the hive. When I tried to lift the inner cover, it snapped open, as if it had been glued shut. This was my first encounter with propolis. Worker bees collect this reddish brown resin by gathering the sap from tree buds. Bees use propolis, or bee glue, as beekeepers call it, to seal up cracks and crevasses in their hive to protect the colony from weather, bacteria, and disease that may enter the hive. Thus, they maintain a sanitary environment. One thing I can tell you is that once propolis gets on your clothing, it will remain there for a lifetime. A wise beekeeper carries a spray bottle of alcohol to the hives to help dissolve propolis that may get stuck to fingers or clothes. Later, I would learn to respect propolis for all its wonderful benefits and gently scrape away the excess for my own personal use in healing remedies.

Once I removed the inner cover, I leaned it against the side of the hive as well. Leaning the covers against the hive, as opposed to laying them flat, ensures that any bees still crawling around on them will not get crushed. Then with both hands, I lifted off the feeder box and placed it on the grass. The bees had completely devoured the entire batch of sugar water, and I would need to refill it before closing up the hive.

I was relieved to observe that my honeybees were calm. They crawled over the tops of the frames, ignoring my presence. I was taught that when the worker bees line up along the top bars of the frames and begin staring at you, it's time to smoke again. But these bees seemed to have no interest in me. I peered inside the hive and saw the queen cage sitting between the two frames where I placed it last weekend. A wave of anticipation came over me as I reached in between the frames to dislodge it. As I removed the cage, along with it came a pristine white piece of *burr comb*. Burr comb is a piece of beeswax honeycomb that the bees make where there is more than ⅜ inch of space anywhere within the hive. This exact measurement is Langstroth's bee space, and it is the perfect size for a worker bee to pass through. And, as Langstroth discovered, if the bees detect this amount of space anywhere in the hive, they build comb to fill it in. Such a space had opened up in my hive because we'd removed the fifth frame to make space for the queen cage. Once the queen cage was removed and the fifth frame set back in its place, there would be no extra space, and the bees would make wax on the frame, as they should.

I saw immediately that the sugar cork had been eaten away entirely, and peering into the tiny cage, I could see no one was left inside. My queen had been released into her castle, and her attendants had joined her. I placed the empty cage, with the burr comb still attached, to the side so as not to damage it. It would be the first of many beekeeping specimens that I would acquire and save.

Now it was time for me to remove a few frames and see if the bees had drawn out the beeswax foundation into cells. Frames number four or six were good choices since they were on either side of where the queen cage had hung. And, it is in the middle of the hive body where the bees generally cluster and the queen begins laying eggs. Using my hive tool, I first gently pried the outermost frame away from the others and placed it on the grass. Then, gently sliding the other frames to the side, I lifted the sixth frame from the hive and held it at a five-degree angle to the sun, just like Mr. B had taught me. I could see that the foundation had been drawn out into beautiful, pure white cells, and they were already filled with a shimmering, clear liquid that was obviously nectar. Bees build the cells at a five-degree angle to the beeswax foundation, which makes it tricky for beekeepers to spot the new eggs. But if you hold the frame at the correct angle to the sun, you can spot the eggs, which should be sitting at the bottom of the cells. Staring closely at those cells and maneuvering the frame till the sun hit it just right, I could finally see the practically microscopic eggs. My queen had begun laying in the center of the frame in the typical oval pattern that I recognized as the brood nest. A single egg was placed inside of each cell, and she had barely missed a cell along the way. This consistency is the sign of a good, full brood pattern. Missing more than a few cells would have created a spotty laying pattern, which could mean a queen that was old or failing.

Above the brood nest was an arc of cells that held a spectrum of colored pollen—yellow, gold, brown, some with a greenish tint, and even red—each representing a distinctive species of plant. Between the pollen and the top bar of the frame were freshly capped cells filled with honey.

Several bees crawled on my sleeves and around the top of the frames without paying me much mind. Their faint buzzing allowed me to remain calm and work quickly. As I was about to replace the space where the queen cage hung with the fifth frame, I noticed a group of bees were *festooning*, an adorable behavior in which the bees hang on to one another, creating a chain across the space between the two frames.

I was sorry to have to insert the missing frame, because when I did, the chain of bees disbanded. I added my feeder box on top and refilled it with more sugar-water solution, being careful not to spill a drop. Sugar left around the hive could invite visitors like hungry raccoons and, in some parts of my state, bears. Now that my inspection was complete, I closed up my hive and let my bees return to work.

It did not take long for me to become comfortable opening up my hive and working with my bees. Before each inspection, I went through the sequence of events in my head before actually proceeding. I discovered this mental review helped me not to forget any important details. Afterward, I always recorded my hive visits and observations in my beekeeper's journal, so I could accurately report any interesting or unusual activity to Mr. B. These records helped me especially when I had a question or concern about the hive or the bees' behavior. I looked forward to each visit as a new adventure in honeybee education and welcomed every opportunity to observe the bees' behavior. On first observation, worker bees seem to have a simple approach to life, yet the complex and harmonious workings inside the hive tell another story. In time, I would be able to read their behavior. For example, it became clear that on bright, sunny days and when there was plenty of pollen and nectar, the bees seemed quite content. On days when there was little food for them to forage, they were defensive. Beekeeping gave me immense pleasure and purpose. And my bees were always teaching me something about nature and life.

Honeybee Communication

In order to communicate information to their sisters, worker bees have developed a sophisticated system of dancing around the hive. A honeybee that returns to the hive will do a dance telling her sisters where sources of nectar, pollen, water, or propolis can be found. These dancing bees are called *scout bees*, and their dances are called the *wagtail dance* and the *round dance*. Through these dances a honeybee can communicate distance, direction, and the quality and quantity of the food source by using the direction of the sun. First, each worker smells and tastes a sample of the prized floral source that the scout has brought back to the hive. Then they observe the scout's dance.

The wagtail dance is preformed on the vertical honeycombs inside the hive, and the steps create a figure-eight pattern. To picture the pattern, first envision the face of a clock. The dance begins at six o'clock with a short walk and wiggle straight up to twelve noon. Then the bee moves to the right, past one, two, three, four, and five o'clock, to reach six again. Another walk and wiggle take the bee up to twelve again, but this time she turns left, moving past eleven, ten, nine, eight, and seven o'clock, and back to six again. The angle of the steps indicate the

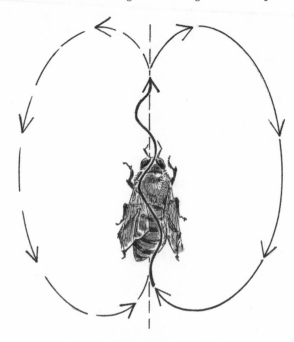

STEPS OF THE WAGTAIL DANCE

location of the food source relative to the sun. If the forager bee dances for a short time, the source is nearby, and if her dance lasts longer than a few seconds, the source could be more than one mile away.

The round dance is a simpler circular dance that is performed for food sources close to the hive, and the more complicated wagtail dance describes more distant sources.

Multiple foragers may be found dancing simultaneously for their fellow workers at the hive. One worker's dance may tell of nectar sources, and another of water sources. The workers at the hive may watch several dances before leaving to find one of these sources. Often the new foragers will not locate the source immediately, but they have the ability to investigate a general area until they locate the specific site. Using their five eyes and their keen sense of smell, channeled through their antennae, these highly intelligent creatures then return to the hive and recruit more foragers. And the dancing continues.

THE GROOMING DANCE

The grooming dance is another form of communication. A honeybee dances to tell another bee that she needs grooming. In this dance, the bee rocks her body from side to side with her legs spread apart. The movement lasts only a few seconds, until another bee begins grooming her to remove pollen or mites.

WASHBOARDING

Washboarding is a somewhat mysterious behavior that occurs at the entrance of the hive. Worker bees stand in groups with their bottoms up and their heads down. Their four hind legs support their tiny bodies while their two front legs scrub the surface of the wooden hive entrance. To me, they really look like they are line dancing. I've heard they are polishing the hive by sealing up cracks in the wood, but no one has seemed to figure out the exact meaning of this behavior.

Swarming: They're Not Running Away, Are They?

Bee swarms are a common event in the spring when the hive population increases and the brood nest becomes overcrowded. Although swarming honeybees may appear (and *sound)* threatening, they do not carry honey, pollen, or young bees in need of protection; therefore, they are not terribly agressive.

When a hive swarms, 30 to 70 percent of the bees leave with their original queen to form another colony, while the remaining bees stay in the original hive and raise a new queen. A few days before swarming, the original queen will begin to lose weight and lay fewer eggs. The bees, noticing these changes, will begin the process of raising one or more new queens inside what beekeepers call *swarm cells*. Such cells are similar to queen cups but are often found protruding from the bottom of the frames and are often even attached to the wood of the frame.

The swarm cells are filled with royal jelly and capped while the queen larvae develop. Before the new queen emerges she will let out a high-pitched, shrilling sound beekeepers call *piping*. This sound is an announcement and warning to any other new queens, and even to the old queen. When more than one queen emerges from the swarm cells they will either fight to the death, with the victor becoming the new queen, or one of the queens will leave with a portion of the hive population in a secondary swarm.

Swarming is a disconcerting experience for the beekeeper, but once you know the signs, it is possible to manage it before it happens. The most obvious sign of an upcoming swarm will be the swarm cells found at the bottom of the hive frames. Beekeepers regularly remove these cells with their hive tool to prevent swarming. During the spring, beekeepers might also reverse the deep bodies to create more room in the hive for brood rearing. The queen tends to begin laying eggs in the bottom deep and then move upward to the top deep. By the time spring arrives she has filled the top deep with eggs, so reversing the deeps will once again create more space for her to move into and lay, thus limiting the feeling of overcrowding in the hive.

Unfortunately, I did not learn about swarming until my first swarm was upon me. It was mid-afternoon during my second spring as a beekeeper. I was working in my garden, transplanting a box of seedlings into their new home of raised beds, when I looked up and saw what looked like a miniature tornado at the entrance of my beehive. Thousands of honeybees hovering in the air like a turbulent storm. I dropped my small spade and cautiously walked over for a closer look. As I approached, I could hear the bees' powerful, low-pitched buzzing. Just as I arrived, the swarm began to move slowly toward a tree branch overhead. Their migration was a dramatic sight to behold—perfectly orchestrated, as if there should have been an operatic aria playing in the background. As the swarm approached the branch, several bees lighted on the limb while others followed. There they united as a huge cluster dangling from the branch. Spectacular! The cluster grew larger and larger, the bees clinging to one another to form what looked like a gob of honey dripping off the end of a spoon. It swayed gracefully in the light breeze. And then, rather abruptly, the swarm fell to the ground in a clump.

My maternal instincts kicked in, and I dropped to my knees to find the queen in the pile of bees that now lay on the grass underneath the branch. Somewhere in the center of that swarm, I knew, was its queen. I thought if I could capture her and return her to the hive, the swarm would soon follow. I found a long twig, which I used to gently spread apart the cluster of bees, and I began searching for the queen. My queen was not marked, so I had to rely on a keen eye to identify her in the pile of tens of thousands of bees. The more bees I spread, the deeper into the cluster I dug. Careful not to hurt a single bee, I patiently continued spreading them out until at last I spotted her. With

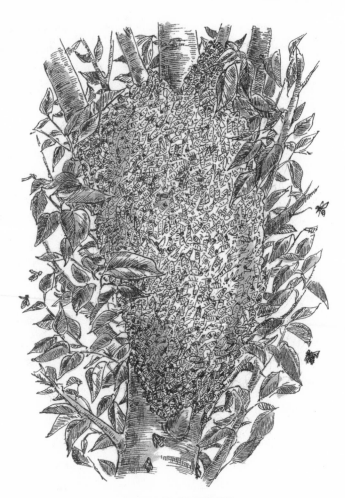

A SWARM OF HONEYBEES

the twig, I guided her away from the cluster and cupped her between my two hands. Although the queen has a stinger, it is not barbed, nor is she ready to use it. I held her loosely in my hands, giving her enough room to crawl.

Unfortunately, as I tried to rise up from the ground, I missed my footing and had to throw out my hands and arms catch my fall. The queen escaped and rose up into the air like a helium balloon, but then, losing momentum, she fell back down to the ground, where I was able to retrieve her again. I walked her back to the hive, where I lifted the outer cover with my elbow and pushed the inner cover off to the side with my knee. I placed the queen back into the hive, where I hoped she would remain. After replacing both covers, I waited and watched. Amazingly, within just a few minutes, the rest of the honeybees slowly returned to their hive. It was incredible to watch. Before dusk, most of the bees had come back to the hive. Little did I know that this was not the end of the swarm.

Throughout the entire next morning, everything within the hive appeared status quo. But come mid-afternoon, at approximately the same time as the day before, I saw the tornado of bees on the move again. They swarmed in a completely new direction and farther and higher than the previous day. And this time I did not have the good fortune to see where they'd landed. I called Mr. B to tell him the sad news.

Mr. B chuckled at my story and told me I had done the right thing trying to capture a swarm, but that you cannot return a queen and half the colony to the same hive. Instead, I should have put the swarm in a new, unoccupied hive. When I had tried to return the queen and her followers to their hive, they swarmed again because that colony had already prepared for this dramatic event by raising a new virgin queen. After further observation, I was able to locate the swarm cells at the bottom of the frame.

Mr. B explained the only real way to prevent it from happening again was to clip one of the queen's wings, so she could not take off. This seemed drastic, though I was assured it did not hurt her in any way. Another thing beekeepers do to keep the queen inside is trap her in a wired plaque, thus preventing her from moving around the hive. I now know to watch for another swarm each spring and have a fresh hive ready for my swarming colonies, just in case.

CHAPTER 7

Beekeeping and Honey Throughout History

Entering the beekeeping world fueled my fascination with bees, hives, and honey. During my explorations—in both travel and reading—I've discovered a wealth of information about the history of beekeeping and the roles bees and honey have played in cultures around the world. Here are just some of the fascinating facts my love of bees has brought to me so far.

Evidence points to the existence of bees 100 million years ago. A piece of petrified amber, found in a mine of northern Myanmar (formerly Burma), shows an almost perfectly preserved honeybee, practically unchanged from how bees look today. This specimen proves that honeybees are one of the oldest creatures remaining since dinosaurs and possibly the reason that fruits, vegetables, and animals still exist today. Further evidence of beekeeping appears in ancient literature. Mention of honey was made in cuneiform, the first system of writing, done on clay tablets in Babylonia and Sumeria. Honey also appears in the ancient Sumerian poem *The Epic of Gilgamesh*, which dates from 3000 BC.

Long before sugar cane and maple syrup were discovered, honey was the first and only sweetener available to ancient man. A drawing discovered in 1921 inside a cave called la Cueva de la Arana (the Spider Cave) in Bicor, Spain, is the earliest known image of a honey hunter. Estimated to be approximately 15,000 years old, it portrays a man using a rope to climb the side of a mountain and carrying a basket in one of his hands. Hovering around him are five honeybees, which appear to be emerging from a hole in the side of the rock. Another man is hanging on to the rope just below him.

The first book about honey was published on 1759 in London. Written in English by a Covant Garden apothecary named Sir John Hill, its rambling title would never make it on the shelves in today: *The Virtues of Honey in Preventing many of the Worst Disorders, and in the certain cure of Several Others...*

Beekeeping and Honey Around the World

Customs and rituals about honey and honeybees weave through every ancient culture in the world. The information and stories have been passed down through the generations both orally and in writing.

I have compiled a brief compendium of international honeybee notes and facts.

EGYPT

- The ancient Egyptians were the first known organized beekeepers. The honeybee was the symbol of Lower Egypt, and keeping bees was a part of everyday life there. The Egyptians had sophisticated knowledge regarding bees. Carvings on temple walls tell stories of harvesting honey from trees and rocks. The Egyptians were also the first migratory beekeepers. They would place their beehives on boats and float them up and down the Nile to pollinate the crops along the river.

- Propolis, beeswax, and pollen played important roles in health, in the creation of various medicines, and in the embalming process. Honey was found in clay pots inside the tombs of the Egyptian pharaohs, who were thought to consume it in the afterlife. Mummified bodies have been perfectly preserved after having been covered with honey, beeswax, and propolis.

- Actual Egyptian medical texts, specifically the Edwin Smith Papyrus and the Ebers Papyrus, both dating from 1550 BC, recognize honey for its topical healing properties for wounds, sores, and skin ulcers. These papyri mention honey as an ingredient in at least nine hundred remedies. Egyptian doctors customarily prescribed honey and milk for the treatment of respiratory ailments and throat irritations.

CHINA

- Beekeeping has been documented as being practiced in China more than three thousand years ago. Oracle inscriptions from the Shang dynasty of the eleventh century BC show bees swarming. King Zhou Wu, the first ruler of the Chinese Zhou dynasty, led his army with a bee flag. He reigned from 1046 BC to 1043 BC.

- Since at least the second century BC, bees have been used in Chinese medicine, including through a technique often called bee acupuncture. During the Ming dynasty (1368–1644), pharmacist Li Shizen practiced traditional Chinese medicine in the Hubei province of China and used it for enhancing yin, or cold energy. Li Shizen wrote in the *Compendium of Materia Medica*, "There are five medicated functions of honey: dispersing heat, supplementing the internal organs, detoxifying, moistening dryness, and relieving pain."

- Another Chinese medical text, *Materia Medica of Shen Nong*, written around 50 BC, says that honey has the ability "to pacify the deficiencies of the five internal organs, [honey] benefits the vital energy and supplements the internal organs; it also relieves pain, detoxifies, treats numerous illnesses and counterpoises hundreds of herbs."

- Huang Gongxiu says in *Probes on Materia Medica*, written around 1736–1796, that "When honey is fresh, it is cool in nature and clears venting. When it is cooked, it is warm in nature and supplements the internal organs. Its flavor is the purest. Whoever feels deficiency in the internal organs, dry and unresolved, uncomfortable and un-balanced; with sudden pain in the heart or the stomach, have coughs and diarrhea, feels dizzying and looks withered, can all use honey to remedy the conditions."

GREECE

- In *Natural History*, Aristotle wrote about his observations of the honeybee, including how it collects the juices from flowers and carries them back to its hive. He recognized that the juices were watery and that it took some time for them to attain the correct consistency of honey. Unfortunately, his observations were not completely accurate; for example, he wrote that honey is deposited from the atmosphere.

- The honeybees' organized structure inside the hive inspired Lycurgus, the founder of the ancient city of Sparta. He was so impressed by the social structure of the hive that he used the bee colony as a model for his government.

- Traces of honey were found in cooking pots from the Mycenaeans. Today, Greek desserts, such as baklava and yogurts, are often slath-ered in honey.

ISRAEL

- What is now thought to be one of the first full-scale apiaries was found in excavations at Tel Rehov in Israel's Beth Shean Valley. The apiary dates from the period of the First Temple (in the tenth to early ninth century BC) and housed approximately two hundred hives. The beehives resemble clay tubes lying on their sides and stacked on top of one another. Remarkably, they resemble the style of apiaries found in Middle Egypt. All indications are that beekeeping and the extraction of honey and honeycomb was a highly developed industry at Tel Rehov at this time. When the Bible called Israel the "land of milk and honey," there is a good chance it was referring to Tel Rehov, one of the most important cities and a beekeeping center.

INDIA

- Honey has an important role in the festival of Madhu Purnima, celebrated by the Buddhists of India. When the Buddha made peace with his disciples, he retreated into the wilderness. There a monkey gave him honey to eat. In honor of the monkey's generosity, Buddhists often give honey to monks during the festival.

- One of the most common drinks in India in the first millennium AD was a special ceremonial brew made from sugar, ghi (clarified butter), curds, herbs, and honey. It was given to guests, to suitors about to ask a young woman for her hand in marriage, and to women who were five months pregnant. It was also used to moisten the lips of a newly born first son. The name for this drink was *madhuparka*. The first two syllables of *madhuparka* mean "honey"; the Sanskrit word *madhu* and the Chinese word *myit* are related to the words *mit* of the Indo-Europeans (Aryans), *medhu* of the Slavs, and *mead* of the English-speaking.

RUSSIA

- Russian author Leo Tolstoy was a beekeeper and uses beekeeping as a metaphor in his novel *War and Peace*. In it he describes the evacuated city of Moscow by saying, "It was deserted as a dying, queenless hive is deserted." His wife, Sonja, later wrote in her diaries of Tolstoy, "The apiary has become the center of his world for him now, and everyone has to be interested in Bees!"

FRANCE

- When Napoleon Bonaparte declared himself emperor of France in 1804, he refused to allow the pope to crown him and instead placed the crown on his own head. Napoleon's coronation robe was decorated with embroidered bees, a symbol, taken from Merovingian kings of the past, of "a republic with a chief, with a sting but producing honey." There is even speculation that the French fleur-de-lis originated as the graphic outline of a bee (taken from the emblem of Childéric, a Merovingian king).

  Louis XII (1462–1515) used a beehive as part of his coat of arms, "but the National Convention rejected this emblem for the Republic, 'because bees do have a queen.'" The bee was a symbol for the French between 1804 to 1814 in the First Empire, during the Cent-Jours (1815), and also in Napoleon III's rule during the Second Empire (1852–1870).

BEESWAX AND HONEY IN HISTORY

- The ancient Egyptians used beeswax to paint their sarcophaguses. The beeswax was heated and mixed with pigments and then applied with a paintbrush to the surface of the coffins. The beeswax created a waterproof color that protected the surface of the stone.
- The Persians embalmed the dead with wax.
- Romans carved death masks and life-sized figures out of beeswax.
- Ancient Egyptians, Assyrians, Greeks, and Chinese, and modern Germans as late as World War I, used honey as an antiseptic for wounds during wartime.

- The Greek physician Hippocrates referred to as the father of medicine, cured skin ailments and ulcers using honey.

HONEYBEES IN ART

Reading accounts of beekeeping in ancient history is intriguing to me. In my own travels to foreign countries I have uncovered more of the history of bees and honey, as well as many examples of honeybees in artwork. On a visit to Vatican City, for instance, I discovered bees and honey depicted in marvelous paintings on crumbling stone walls and decorative patterns of honeybees adorning the inner chambers of sanctuaries. The honeybee repeatedly appears as an icon throughout Roman, Florentine, and Venetian art.

In the seventeenth century the famous sculptor Lorenzo Bernini was commissioned by Pope Urban VIII (or Urbano Barberini) to sculpt his family crest. The final sculpture, completed in 1644 in Rome, was called *Fontana delle Api,* or the Fountain of the Bees. The basin is in the shape of an open clamshell. Inside the shell stands the Barberini family crest, adorned with three honeybees. But why bees, you may ask? Because they symbolize diligence (the hard work of the worker bees), social organization (hierarchy within the hive), and purity (production of beeswax, the purest form of candle wax). Around the mid-1800s the fountain was dismantled, and some parts were lost. In 1915, it was restored and moved to the north side of Rome's Piazza Barberini, where the piazza meets the Via Veneto.

At the Palazzo Barberini, I was humbled by it's best example of honeybee art: a huge fresco, Pietro da Cortona's masterpiece, entitled *The Triumph of Divine Providence,* which fills the ceiling of the grand salon. This baroque painting was begun in 1633 and was completed in 1639. It celebrates the spiritual and secular power of the Barberini family's glory. The highly detailed painting gives the illusion of figures floating above the room and is full of various symbols, including honeybees, which bob among the figures.

In Slovenia, during the eighteenth and nineteenth centuries, beekeeping used to be an important element in people's lives and an important line of business. Today, the Radovljica Museum of Beekeeping in the town of Radovljica, Slovenia, boasts a fine permanent exhibition of painted beehive panels. These wooden panels depict colorful stories of the life of the honeybee and were hung on the sides of beehives. The traditional art of painting beehive panels flourished in the eighteenth and nineteenth centuries among Slovenian farmers, who at the time made up the country's largest social class. Together with Slovenian folk songs, legends, fairy tales, and the remarkable creations of traditional Slovenian architecture, beehive panels represented this culture's limitless abundance of folk imagination, thought, and creative expression.

Honey in Culinary Recipes

Nearly every culture uses honey in its cuisine. Here are examples—some familiar and some rare—of ways honey has been enjoyed throughout history.

AMBROSIA: In Greek mythology, the Olympian gods preserved their immortality by eating honey. It was accompanied by nectar, wine of the gods.

APICIUS ROMAN CHEESECAKE: Apicius was a first-century nobleman who has been credited with penning the very first recipes. His love of food and cuisine is reflected in his 468 recipes. Honey was one of his ten favorite ingredients and he used it as a preservative, a condiment, and an ingredient in making wine. I credit Apicius as being the first to pair cheese and honey in his recipe titled "Homemade Sweets and Honeyed Cheeses." When he lost part of his fortune and was no longer able to dine in the style to which he had

Modern Roman Libum Recipe

Reprinted from *A Taste of Ancient Rome*

SERVES: 4

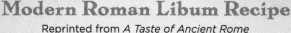

INGREDIENTS:

1 cup all-purpose flour
8 ounces whole ricotta cheese
1 large egg, beaten
1 bay leaf
½ cup Red Bee® wildflower honey

Preheat oven to 425°F. Grease a baking tray. Sift the flour into a medium-sized bowl. In a second medium-sized bowl beat the cheese until it's soft, and stir it into the flour along with the egg. Form a soft ball, and divide it into four parts. Mold each quarter into a bun and place them on the baking tray on top of a fresh bay leaf. Cover with your brick* and bake for 35 to 40 minutes until golden brown. Warm the honey in a shallow dish and place the warm cakes in it so that they absorb the honey. Allow to stand 30 minutes before serving.

*The Romans often covered their cooking food with a domed earthenware cover called a testo. You can use an overturned shallow clay pot, a metal bowl, or casserole dish as a brick.

grown accustomed, Apicius committed suicide by poisoning himself. The word *apicius* loosely translates to the modern word *epicure*, referring to a gourmand.

Libum, a traditional Roman sweet-cheese bread that Apicius wrote about, is similar to our modern-day cheesecake. Making it was often a sacred act and part of the worship of household deities. Served hot or cold, libum should be generously drizzled with honey.

MEAD (OR HONEY WINE): Mead, a fermented beverage made with honey, is the first known alcoholic drink and is known in mythology as the drink of the gods. It is found in every ancient culture and is now experiencing a reemergence in the United States. *Metheglin* is mead made with spices (such as cloves, cinnamon, or nutmeg) or herbs (such as oregano, hops, or even lavender or chamomile). Fermented mead with grape juice is called *pyment*. Mead made with berries, like strawberries, blackberries, or raspberries, is called *melomel*. The list goes on and on, since honey wine pairs well with all types of ingredients to create a truly unique drink.

Every country has its own version of distilled alcoholic beverages made with honey. Some popular ones are Benedictine in France, Drambuie in Scotland, Irish Mist in Ireland, Grappa al miele in Italy, Krupnik in Poland, and Barenfang in Germany.

Telling the Bees

There is an old beekeeping tradition known as "telling the bees." First, the bees must know everything that goes on in their keeper's family, including births, deaths, illnesses, and marriages. Then, upon the death of the beekeeper, a close family member should approach the hive, knock three times with the key to the house, and whisper the news to the bees. The bees, it is said, need to be assured that someone will take care of them after their keeper has died; otherwise, they will abscond or not produce honey. Two nineteenth-century New England poets, John Greenleaf Whittier and Eugene Field, wrote poems entitled "Telling the Bees."

Honey in Love

- The term *honeymoon* originated with the Norse practice of consuming large quantities of mead during the first month of a marriage. This practice was believed to ensure fertility during the lunar cycle.
- In early Greece and Rome honey symbolized fertility, love, and beauty. Greek mythology says that Cupid dipped his arrows in honey to fill a lover's heart with sweetness.
- In Greece it is customary for a bride to dip her finger into honey and make the sign of the cross before entering her new home. This gesture would bring her a sweet married life and good relationship with her mother-in-law.

Mad Honey

There is some honey known to be toxic to humans. Popularly known as *mad honey* or *crazy honey*, it is most commonly found in the northern hemisphere and produced from the spring flowers of rhododendrons, mountain laurels, oleander, and azaleas. The beautiful and fragrant yellow jessamine that turns the Southern swamps to gold in the springtime also has the reputation of yielding poisonous honey. The nectar of these plants may contain grayanotoxin, a compound that is both psychoactive and poisonous to humans, but harmless to bees. However, during the time when azaleas and rhododendrons bloom, other flowers more appealing to the honeybees are usually available. Also, the shape of the azalea flower makes access to nectar difficult for honeybees. Azalea and rhododendron honey remains toxic for only a very short period, so humans rarely encounter lethal honey. In fact, because honeybees act as biological filtering agents, any toxic levels of chemicals brought back to the hive (in pollen, nectar, and propolis) would kill the bees and their brood before even reaching a human.

Nevertheless, the effects of mad honey have been reported in Western literature as early as 401 BC, when Greek general Xenophon, a follower of Socrates, led 10,000 Greek soldiers into battle. Setting up camp in an area named Colchis in Persia, they landed upon wild bees and proceeded to eat the honey inside the hives. General Xenophon described the effects of toxic honey in his history *Anabasis Book viii 18–23*: "All the soldiers who ate of the honeycombs lost their senses, and were seized with vomiting and purging, none of them being able to stand on their legs. Those who ate but a little were like men very drunk, and those who ate much, like madmen and some like dying persons. In this condition great numbers lay on the ground, as if there had been a defeat, and the sorrow was general. The next day none of them died, but recovered their senses about the same hour they were seized. And the third day they got up as if they had taken a strong potion."

Mad honey was used as a weapon in 67 BC, when Roman general Pompey led an invasion to conquer King Mithridates of Pontus in the Trebizon region of the Black Sea. The Roman army had forced Mithridates's men to retreat. But Mithridates's men then left out jars of rhododendron honey as a peace offering to the Roman army. When the Romans feasted on it, they fell into intoxicated fits, and Mithridates's army was then able to defeat them. This incident is the first recorded use of honey as a weapon.

Honeybees in Religion

- The word *honey* appears fifty-six times in the King James version of the Bible.
- The bee and the hive have long been symbols of industry and regeneration, wisdom and obedience, and have places in Egyptian, Roman, and Christian symbolism. The hive is often seen in Masonic illustrations of the eighteenth and nineteenth centuries.

- Throughout history, abbeys and monasteries were centers of bee-keeping, because beeswax was prized for candles used in churches.

- Saint Ambrose, bishop of Milan from 374 to 397, is the patron saint of beekeepers, and statues made in his honor always include a bee-hive. It is said that when he was a baby, a swarm of honeybees settled upon his face. The bees left without stinging him, but they did leave behind a drop of honey. His father believed this incident was a sign that his son was destined to be a sweet-tongued preacher.

- The prophet Mohammed is quoted as saying, "Honey is a remedy for every illness, and the Qur'an is a remedy for all illnesses of the mind. Therefore I recommend to you both remedies, the Qur'an and honey."

- During Rosh Hashanah, it is traditional to eat apples dipped in honey to symbolize the hope for a "sweet" new year. The apple is dipped in honey, the blessing for eating tree fruits is recited, the apple is tasted, and then the apples and honey prayer is recited.

Bees in Currency

- In ancient Rome, honey was highly valued and considered somewhat of a luxury good, reserved for only emperors and the wealthy. This natural sweetener was so valuable that it was common for the Romans, like the Egyptians and the Aztecs of Central America, to pay their taxes in honey.

- The familiar icons of the honeybees and the skep have appeared on coins and postage stamps as far back as the sixth century. Maybe this is why honey is frequently referred to as liquid gold!

CHAPTER 8

From Hive to Home: Making and Harvesting Honey

The honey-making process inside the hive is a perfectly orchestrated symphony of events. And it is because honeybees hoard their honey and other resources that beekeepers are rewarded with excess honey.

The worker bee begins foraging for nectar in the first three to four weeks of her life. She gathers up nectar by visiting flowers within two to three miles of the hive. She sucks up the nectar from many flowers with her long, tubelike tongue, or proboscis, and then stores it in her special honey-sac stomach. This stomach is separate from her digestive stomach and holds up to 70 mg of nectar, almost as much as she weighs. A honeybee must visit fifteen hundred flowers to gather enough nectar to fill up her honey stomach. She then carries the nectar back to the hive and turns it over to the *house bees*.

A younger house bee accepts the nectar from the foraging bee by sucking it out of her honey sac through her mouth. These house bees mix the nectar with their own enzyme, called *invertase*, which breaks down the sugary nectar, or sucrose, into glucose and fructose, making it possible for the bees to digest. The nectar is then placed in the honeycomb cells. Worker bees inside the hive fan the liquid nectar with

A WORKER BEE GATHERING POLLEN AND NECTAR

their wings, which helps to evaporate the extra moisture and bring the water content of the nectar down to 18 percent. This process ripens the nectar into honey. Honey harvested before it is ripened, with a moisture content higher than 18 percent, can cause the naturally occurring yeast cells in honey to ferment. Fermented honey tastes a bit like vinegar and is exceedingly runny. Somehow the bees instinctively know when the nectar is at the correct moisture content and the cells containing it are ready to be capped with beeswax. Once the bees have filled all the cells within a frame with honey, they cover the cells with pure-beeswax cappings, creating what I call the beautiful stained-glass-window effect.

A HOUSE BEE ACCEPTING NECTAR FROM A FORAGING BEE

Honey Shallows

During the first year of beekeeping, your bees will be busy drawing out the hive's twenty frames of beeswax foundation into cells for brood rearing and honey and pollen storage. The worker bees excrete wax from their glands and use it to build cells on the foundation inside each frame. Once the beeswax cells are drawn out, the queen will start laying eggs in the center of the frame as the worker bees begin filling up others at the top with nectar for making honey. As the queen lays eggs, the colony will grow in number, building a strong population of worker bees. An enterprising colony of honeybees will make far more honey than it needs to survive. Honeybees will naturally continue making honey as nectar and space are available. You do not always have control over the nectar availability, but you can control the space and give your bees extra room to make honey.

You should never remove honey from the hive's two deeps—the two main boxes are where the bees live and raise brood—because that's the honey the bees eat and feed to their larvae. Instead, you place a third box, called a *shallow super*, on top of the hive body, for the bees to fill up with excess honey. You can then harvest the honey from this extra space without disturbing the main hive body. This box is called *shallow* because it is shallower in height than the hive body where the queen lays her eggs and brood is raised, and *super* because it goes on top, or *superior*, to the deep boxes. These boxes are also sometimes referred to as *honey supers*.

Inside a shallow super there are nine frames, instead of the typical ten found in the deeps. Each frame is 5⅜ inches deep and holds a piece of beeswax foundation similar to the frames inside the hive body. Having nine frames, rather than ten, gives the bees more space in between the frames to build up the wax cells, allowing them to build the cells to a maximum depth.

To prevent the queen from walking up to the honey shallow to lay her eggs, there is a piece of equipment called a *queen excluder*. This flat wooden frame has a grid in the center that acts like a fence; it has just enough space for the smaller worker bees to pass through and carry their nectar for honey making. The chubbier drones will not fit through the queen excluder, either. If you choose to use a queen excluder, place it on top of the two brood chambers with the shallow super on top. You put your hive's inner cover and outer cover over the shallow, just as you would normally put them over the hive to close it. From the outside, your hive will look a bit taller with these accessories.

Placing these shallow supers on your hive at the appropriate time is familiarly called *supering*. The best time to put a honey shallow on your hives is early spring, which in my area, the Northeast, is considered May or June, when the first flowers are popping up and the nectar flow is in full swing. The earlier you place your shallows on your hives, the better chance you have of getting those workers to draw out the beeswax foundation and fill the cells with honey. Early supering gives your colony plenty of time to complete all their honey-making duties in time for your honey harvest in the fall.

By September, if you had a good summer and the weather cooperated, the bees will have filled all nine frames inside the shallow with honey. If the shallow becomes full or looks crowded, and your bees are still bringing in nectar, it is time to add another shallow on top of the stack. The ideal productive honey season is one where your hive has a large, healthy population of foraging worker bees. This part depends upon your queen and her ability to reproduce. The weather also plays a role in honey production. A pleasantly warm season with occasional rainfall is amicable to pollen production. The bees also need to have continuously blooming nectar-bearing flowers to forage on. A summer that is too humid makes the flowers wilt and dry up the nectar, and one too rainy will wash away the pollen.

Once the worker bees have completely filled each hexagonal cell with honey and placed a wax cap over it, you should remove the honey-filled frames from the shallows. It is important to make sure that the bees cap at least 80 percent of the honey-filled cells on each side of a frame before you remove it. Honey from uncapped cells is watery and not ready to harvest, because the bees have not yet finished evaporating the water content of the nectar to 18 percent. If you harvest honey from frames that are not fully capped, you run the risk of having honey that will ferment in the bottle.

Beekeepers remove their honey shallows and extract their honey for bottling at different times, depending on when fall arrives in their location. In the Northeast, this is usually late August and September. Although you remove the honey shallow, the main two deeps still contain enough honey for your bees to carry on the colony throughout the rest of the fall and winter. Removing my first honey shallow from my own hive was a memorable occasion. The sight of a frame filled with honey, perfectly capped and glinting in the sun, is a beautiful and gratifying thing.

Harvesting the Honey

Before the honey can be harvested, the shallows must be removed from the hive, and the honeybees must be removed from the shallows and the frames. There are two ways of removing the bees from the shallow. One is by using something called a *bee escape*. A bee escape is a plastic insert that goes into that oval hole in the center of your inner cover and acts as a one-way door. It allows the bees inside the honey shallow to crawl down into the deeps, but not return to the honey shallows. You place the bee escape between the shallow and deeps a day or two before you want to remove the shallow from the hive. Generally, after twenty-four hours, all of the bees will have left the shallow and will not be unable to return, allowing you to remove the honey super and leave the bees behind.

The second method of removing bees is for the impatient beekeeper, and it is the one that I prefer. It involves a *fume board* and some aromatic spray. A fume board is a flat board resembling a hive cover with foam lining the inside. Essentially, you remove the inner and outer covers of the hive and replace them with the fume board. But before you do, you spray the fume board with an aromatic spray that smells like almonds. This spray clears the bees out of the honey shallows. Bees dislike the aroma so intensely that they will move down into the deeps and away from the fume board, thus abandoning the shallow and its frames almost immediately. After installing the fume board, you need to wait only about twenty minutes before removing the shallow.

When you are ready to remove your honey shallow, use your hive tool to gently pry it away from the hive's top deep, leaving the fume board on top of the shallow. Check the shallow to make sure there aren't any live honeybees left inside. You can brush away any few bees left behind using a simple household brush. Store the shallow in a plastic bin with an airtight lid that snaps shut, to protect it from mice and ants that love to feast on fresh honey.

Honeybees can get defensive while you remove the honey shallows, so you may want to wear your protective beekeeper's veil and gloves. Remember that in the late summer, when you are most likely removing your honey, there may not be many nectar and pollen-producing plants for your bees to feed on, so they may defend their honey more fiercely. Also, be prepared to deal with lots of propolis, which can make removing hive parts a challenge.

When you have removed the honey-filled shallow, it is time to begin extracting the honey. Extracting and bottling your first honey harvest is a glorious and sticky event. Give yourself and any helpers you may be able to enlist plenty of time to do the job. Confining the job to just one room will help you contain the big, sticky mess that accompanies the process. A warm room makes the job easier, because cold honey can be difficult to extract. Spreading a plastic drop cloth or tarp under your work area will save you time on the cleanup afterward. A large table allows you to spread out your equipment. Cover it with plastic or newspapers to protect it. Keep a bucket of warm water and a few old rags handy just to wipe off your hands and clean up any drips along the way. You will be happy you took these few extra steps to prepare your work area.

There are a few pieces of equipment available for extracting your honey. But if you simply want comb honey, you do not need any fancy equipment. You can slice the honeycomb right out of the frame. A warm knife will work well, or a metal cutter, resembling a cookie cutter and made especially for this purpose, will help you cut a perfect square of honeycomb from your frames. Beekeeping-supply companies sell various types of clear boxes made for packaging honeycomb, or you can keep the comb in a dish in your kitchen, so the honey in it is ready for everyday use.

To extract liquid honey from the comb, you will need a few more pieces of equipment: a *heated uncapping knife*, a *capping scratcher*, an *uncapping tank*, an *extractor* (or *spinner*), a mesh strainer, and a clean five-gallon bucket with an opening called a gate.

The heated knife is used to slice off, or uncap, the wax cappings from the honeycomb, thus exposing the fresh honey inside each cell. You hold the frame in a vertical position over the uncapping tank, resting it on the horizontal bar across the top of the tank, and tilt the frame toward your body at a forty-five-degree angle. Starting toward the bottom of the frame, run the heated knife over the surface of beeswax towards the top of the frame. Be careful to keep your fingers away from the edge of the knife. The heat from the knife will melt the

TOOLS FOR EXTRACTING HONEY

beeswax slightly and make the cappings easier to remove. The beeswax and any excess honey will fall into the tank. A wire screen, positioned inside the uncapping tank, will catch most of the wax while allowing the honey to dribble through to the bottom of the tank. Use a fork to gently scratch open any cells that your knife missed, and continue letting any excess honey drip into the tank.

Once both sides of the frame are uncapped, the frame is ready to be placed into the extractor. This large, stainless steel circular tank spins the remaining honey out of the cells. The frames sit securely inside the extractor, held in place by wire grids. My spinner holds nine frames, or one complete shallow's worth. When the extractor is full, close the lid, grab the crank on the top, and begin turning it. The centrifugal force will pull out a good portion of the honey from the

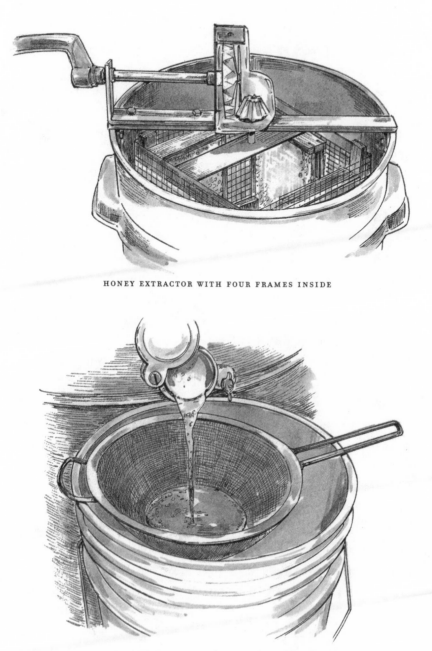

HONEY EXTRACTOR WITH FOUR FRAMES INSIDE

STRAINING HONEY OUT OF THE EXTRACTOR

frames and should also leave the beeswax combs intact. This process is called extracting or spinning your honey. Honey will eventually slide down the sides of the extractor tank and accumulate on the bottom, where it can be dispensed through a tap or *honey gate*. Place your plastic bucket under the honey gate, and place a mesh strainer on top of the bucket to remove the last pieces of beeswax or bee parts. After your bucket is full, you can simply transfer the honey into jars through the bucket's gate.

Any beeswax and honey left in the uncapping tank can also be strained into the same plastic bucket. The pristine white beeswax that caps the cells is the most desirable and cleanest wax. Later it can be melted and cleaned for making candles or beeswax salves.

As for the empty frames of honeycomb, they can be put right back into the hive, where the bees will finish off any remaining honey left on the wax. They will eat it as fast as they can, cleaning up all the drips and leaving you with a perfectly clean shallow super of drawn-out beeswax frames, ready to be placed on your hives next spring.

• • •

ONE OF THE GREAT THINGS ABOUT BELONGING to a bee club is that when one member is extracting his or her own honey, you will often be invited to help and to even bring over your own honey-filled shallows for extraction. Sharing equipment saves time and money, and sharing the work can be a lot of fun. I helped Mr. B harvest his honey in the summer of my first year of beekeeping. Members of the Back Yard Beekeepers Association joined in, and their children and dogs licked up the honey drips as we worked.

One of the most gratifying aspects of extracting and bottling my own honey was being able to label the honey as my very own. No sting or sticky mess could take away from the triumph of seeing my own honey bottled with my Red Bee label. After much research and trial and error, I had developed my signature honey bottle and designed my own Red Bee label. I tried my hand at selling the bottles at local health-food shops and gourmet-food stores, but farmers' markets were where I found the real demand. The appetite for local, handmade products has grown over the last several years, and consumers are beginning to understand the great benefits of pure honey. We have arrived in a world where the respect for artisanal products is now beginning to outshine mass-marketed products. This fact was a truly victory for this former worker bee who longed to be queen.

CHAPTER 9

A Master Class in Beekeeping

Each member of the Back Yard Beekeepers Association is proud of their own apiary, whether small or grand, and enjoys showing off their gardens and honeybees to fellow beekeepers. During my first year of beekeeping there were many occasions to visit other seasoned beekeepers and even help with their hive duties. Early on in my beekeeping apprenticeship, I was personally invited to spend a full day with a member of our club. I earnestly accepted, knowing that this experience would be valuable. So with my beekeeping veil and jacket in hand, I left the house at 8:30 a.m. to begin work punctually at 9.

An active member of our bee club, William was a stocky and serious man who had been beekeeping for many years. Because he was always impeccably dressed in a full suit and tie, complete with cuff links and a tiepin that gave him an old-world demeanor, it was almost unimaginable to me that William could dress down and actually get his hands dirty enough to keep bees. And yet, his backyard, his own private sanctuary, was home to some thirty hives, as well as rabbits, wild birds, and ducks. The yard sat on the edge of a sparkling, manmade pond, both of which made for a breathtaking view from his home. William's wife, Margie, was a dark-haired, quiet woman, and although she did not take care of the honeybees, she made wonderful crafts and products using the beeswax her husband gathered from their hives.

When I arrived at their house on typical clammy summer day in the middle of August, little did I know that we would be opening and inspecting every single one of William's thirty beehives! August, in our region, is the time to evaluate the hives and begin preparing them for winter. When I arrived, Margie greeted me at the door and led me through the air-conditioned house to the backyard. Along the way I spied all types of bee crafts and collectibles. Various bee-themed needlepoint pillows, honeypots, and tiny bee statuettes embellished their home.

"Good morning," I greeted William from their back porch. William peeked his head out from inside his shed, which served as a honey house. "Are you ready for a full day in the beeyard?" he asked. I smiled and followed him around the other side of the pond to a hidden trail that led through the woods and into the beeyard. There, among the trees, sat William's prized honeybee hives. He told me he had visited beekeepers in other countries and learned some new techniques of hive management not taught here in the United States. Each of his wooden beehives looked remarkably different from my own beehive. These were not simple Langstroth-style hives. Custom made and imported from Europe, each was architecturally engineered with unique features, like ventilation chambers and two brood boxes. William told me they were called Kerkhof hives. Some were fancy, with two or four entrances on all sides. They looked similar to a honeybee condominium. I appreciated the unique designs, and the idea of having highly stylized beehives in my own yard was very appealing.

Every beekeeper has their own tricks of the trade, and I soon learned there was never only one way to do things in the beeyard. William donned his veil and announced we were going into the first hive. He did not use the typical smoker and fire to puff his bees, but instead used a squirt bottle of brown liquid known as *wet smoke*. Opening the cover of the first hive, William gave a single squirt to the top of the exposed frames. The liquid smelled something like barbecue sauce, and it did the trick, same as the traditional smoker. Apparently, a little wet smoke goes a long way. Using this handy liquid was quicker than lighting and keeping the flame going inside a smoker, and since we had more than just a few hives to visit, using it was one way to save time.

William then asked me to take out the first frame and tell him what was going on in this hive. I could see that this was a little test for me, and I rose to the occasion. Confidently, I began to pry up the first frame with the end of my hive tool. It was a challenge, as the frame was thoroughly glazed over with beeswax and propolis, which was typical for this time of year. William told me he harvested every single nugget of propolis for making into alcohol-based tinctures. So we scraped and scraped the fragrant, gummy propolis into a jar filled with alcohol, which would slowly melt the propolis down into a liquid state and make it usable. After letting me make a few attempts to lift the sticky frame from the hive, William lent a hand and began prying one side of the frame up with his hive tool. I did the same on my side. All this shimmying made the bees a bit defensive, but I took a deep breath and remained focused on the task. Up came the frame, dripping with brilliant honey and burr comb—a lovely waxy mess. Buzzing contently while I grabbed the whole frame from William, the bees licked the honey from the cells that had been punctured open as the frame made its way out. I held the frame properly, by the corners and at a five-degree angle to the sun. There was capped honey along the top of the frame and much more in the top corners. Near the honey was brightly colored bee pollen, packed neatly inside other wax cells. Tacky propolis lined the sides of the frame, put there by the bees as insulation in preparation for the future cold weather.

I was able to point out the difference between capped brood cells of the worker bees in the center of the frame and the cells of the drones, with their distinctive raised-wax cappings, at the bottom of the frame. In between I spotted tiny eggs at the very bottom of a few cells; they were floating in a clear liquid that, William told me, was the royal jelly. A few cells had O-shaped larvae in them and soon would be capped over by worker bees. William smiled. This hive appeared to be healthy.

William then asked me to look for the queen. This would involve removing many frames and inspecting them on both sides, and also possibly removing the top deep box to check the frames below. I had removed the top deep in my own hive, so I was aware that the deeps could be heavy. A hive in late summer would be filled to the brim with honeybees, honey, pollen, and brood, making the deeps a challenge to lift. Beekeepers have to stay in shape to be lugging around such heavy equipment.

One by one, I began removing and inspecting each frame for the queen. This was no easy task. What a beekeeper really needs to find the queen is a keen eye and knowledge of the inner workings of the colony. It also helps to be orderly during such an inspection. First, I checked the center frames before moving on to the outer ones. I systematically moved my eyes across each frame in an organized pattern, without missing a section. Then I checked for circles of attendant bees, who would be surrounding the queen, most likely in the center of the frame. Luckily, I spotted the queen without having to remove the top deep box. She was placing her behind into a beeswax cell, contentedly laying eggs. Carefully, I returned the frame to the hive so as not to crush her by accident. Then I returned the rest of the frames that I had removed, and we closed up that hive and moved on to the next one.

Each hive had its own distinctive personality. In some the bees were defensive, and in others they were quite docile. The hives sitting directly in the sun were bursting with honeybees fanning their wings at the entrance, working to keep the hive cool. The hives closest to the flower beds had very little activity, since most bees were out foraging. When we reached the seventh hive, William pointed out that it had very little activity at the entrance and scarcely any bees bringing in pollen. And we noticed that some of the drones at the entrance had deformed wings and could not fly. How utterly sad this was—helpless bees who could not fly. These details were reasons to take a closer look. Inside the hive I saw that there were more than a few empty cells among the capped worker brood. William told me that when the queen does not lay an egg in every cell, but leaves large areas open, the result is called a *spotty laying pattern*. Other cells looked as if the worker bees had destroyed them.

Then William asked me to do something unusual: he told me to poke open a capped drone cell with my hive tool and expose the unborn bee. Inside we discovered a white larva with strange reddish spots. The spotted larva was a sign of the all-too-common Varroa mites. These minute parasites cling to drone larvae inside their cells, feeding on their blood. Taking a closer look, we could see that there were also mites visibly clinging to the adult bees inside the hive.

Unfortunately, it was already late summer and this colony's ability to survive the winter was severely compromised. By the time a colony is found to have disfigured bees, it is, more than likely, doomed.

That morning, I learned that a large part of keeping honeybees was preventing and diagnosing diseases. If hives are left unchecked, their honeybees can spread diseases and pests to other hives in the vicinity.

By the time William announced that it was time to break for lunch, the sun was directly above our heads, and the humidity had become unbearable. It had to be close to noon, and I was relieved to take a break. Working our way through so many hives in the middle of August was brutal. I began to appreciate the hard work and long days put in by the migratory beekeepers who work thousands of hives for pollination services. As we walked back to the house, we removed our veils, and I could see the sweat beading on William's forehead. I admired his dedication to his honeybees. This was a full-time hobby for William, and he was a devoted beekeeper.

Margie had prepared us a perfect feast, and we all sat down and dug in. I was parched and starving. For dessert, we dined on homemade yogurt mixed with fresh honey from their hives. Elegant and simple, honey makes everything special. I could see how proud they were serving this traditional combination, and I indulged them by asking for another helping.

We discussed the events of the morning with Margie, who, surprisingly, was familiar with every single one of their hives, although she did not tend to the honeybees herself. After lunch, she invited me into her workroom to show me her handiwork. Fantastic beeswax candles shaped into honey bears and bee skeps lined the shelves. Along with candles, she mixed up luscious lip balms and hand salves using pure beeswax. I was treated to a sampling. The salves and balms were creamy and smooth, and I had never experienced a purer, more emollient moisturizer. After applying them, my skin felt like silk. The bouquet of beeswax and honey permeated the room. The idea of making honeybee products was right up my alley, and thoughts of making pure beeswax skin-care items appealed to my love of natural products. I was so delighted that I knew I would have to try making my own beeswax products at home. Once again I was inspired by the honeybee.

Now refreshed, I was ready for more work with the bees and headed back out to the yard, where William and I continued visiting the rest of the thirty hives. Most were healthy, and we removed the last of his honey shallows and prepared the hives for winter. I would learn much more during the day's hive visits besides how to identify Varroa mites.

Bee Pests and Diseases

There are several pests that can ravage your honeybees and even destroy your equipment. Mice, skunk, opossums, and bears are found in backyards and in most rural neighborhoods. In winter, while bees are clustered, mice will move right into a warm, honey-filled hive, eating the honey and burrowing holes into the wax frames. Skunk and opossums, attracted by honey or sugar-water drippings, will scratch the hive and annoy bees, making the bees nasty and defensive, while bears can totally devastate a complete beeyard in one fell swoop. Hungry bears looking for a midnight snack topple the hives and expose frames, which the bears then rip apart. Attracted by the smell of fresh honey, bears will also feast on bee brood for the protein. Bears have caused irreversible devastation to many beeyards and create many financial and emotional woes for beekeepers.

Honeybee diseases and parasites are all too common visitors to the beehive. Beekeepers quickly learn how to recognize, prevent, and treat signs of problems within their own hives, and many bee club meetings are dedicated to educating beekeepers. Each beekeeper has his or her own practices of monitoring and treating bee maladies. "Ask ten beekeepers how to do something, and you'll get eleven answers," is how I explain beekeeping to folks. There are approved medications available through bee suppliers, but many beekeepers are looking for alternatives to chemicals. Beekeeping clubs, bee suppliers, state bee inspectors, and state agricultural departments are resources where beekeepers can get advice on the care and health of their bees. After all, honeybees are considered agricultural animals that need to be treated as such. Here is a brief list of the most common diseases and pests a beekeeper might have to contend with.

VARROA MITES: Known formally as *Varroa destructor*, the Varroa mite arrived in the United States in 1987, and today mites are found all over the world. Female mites are parasites that look like tiny red-brown ticks. They crawl into uncapped cells inside the brood nest to feed on the blood of a new larva; they prefer drone larvae because of the longer incubation period. Once the cell is capped, female Varroa lay eggs that cling to the larva and then leave the cell with the developed bee.

Advanced signs of a Varroa infestation include mites on drone larvae, bees with deformed wings, and a decline in colony population.

There are many approved treatments such as medications and supplements as well as natural products in the form of liquids, fumes, and powders available through beekeeping suppliers that can be used to combat mites. These natural products are less offensive to the bees and the environment; I like to call them soft chemicals. Beekeepers can also control mite populations within their hives by working with the biology and behavior of certain pests, using a system that is referred to as Integrated Pest Management (IPM). This is a system for identifying and managing pests in an environmentally sensitive way rather than trying to simply eliminate them with toxic methods.

One IMP method you can use to detect mites is a sugar shake. Collect about half a cup of bees—that's approximately three hundred bees—in a small glass jar. Add two tablespoons of powdered sugar to the jar. Shake the jar around, coating the bees with the sugar. Then shake out the bees onto a sheet of white paper, where you will be able to see the mites. You can then return the bees to the hive. This process does not harm the bees; it only shakes them up a bit, and they will lick themselves clean. You can try using this sugar-shake technique in spring and fall. If you have more than three or four mites in spring or ten to twelve in fall, you should think about treating your hive.

Many beekeepers also opt for a bottom board with a screen. When dead Varroa mites drop off the bees, they fall through the screen onto a piece of sticky white cardboard. The beekeeper can then easily count the mites on the white board to monitor mite levels and determine the threshold levels for treatment. Fifty or more mites means you should probably look into treating your bees.

The drone-trapping method involves placing a frame inside your hive with *drone foundation*, a sheet of beeswax with larger cells to accommodate the drones. The queen will lay unfertilized eggs inside these larger cells. Since mites prefer the longer incubation period of the drone larvae, they will naturally be drawn to the frame. Once the frame is covered in capped drone brood, the beekeeper removes the frame and freezes it overnight. Unfortunately, this method kills the drones, but it also removes a large proportion of mites.

TRACHEAL MITES: These microscopic parasites (*Acarapis woodi*) live inside the breathing tubes of queens, workers, and drones alike, and, like Varroa mites, feed on bee blood. Their presence inside the bee causes internal wounds that lead to other infections, makes it difficult for the bee to breathe, and interferes with its ability to fly. Symptoms are *K-wing syndrome*, where the two wings on one side separate so that they look like shape of the letter K, and bees that do not form a cluster in colder temperatures. Often these symptoms are not easily detectable and there is no sure way to know if your colony is infected, so management is the key. Beekeeping suppliers offer a variety of treatments for tracheal mites. I have heard that some beekeepers in Italy place fresh sprigs of thyme, peppermint, and eucalyptus, herbs known to aid the respiratory system, on top of the frames inside their hives.

Feeding your honeybees grease patties is helpful to control tracheal mites in the colony. You may mix the recipe for grease patties, on the next page, at home for your bees.

NOSEMA DISEASE: This disease, caused by the spore-forming parasites *Nosema apis* and *Nosema ceranae*, affects bees' digestive systems. During long periods of cold weather, bees cannot always leave the hive to relieve themselves. When they defecate inside the hive, nosema can spread quickly. Some signs of nosema can be bee feces inside or on the outside walls of the hive. Medication approved for nosema should be given.

WAX MOTH: These female moths (*Galleria mellonella*) can find their way into a hive during a summer night by slipping past the guard

Recipe for Grease Patties

MAKES 8–10 PATTIES

INGREDIENTS:

1 cup solid vegetable shortening

2 cups granulated sugar

1 oz. peppermint or wintergreen essential oil

Over medium heat in a large pan, melt vegetable shortening. When the shortening turns clear, add the sugar. Mix well until the sugar is completely dissolved. Turn off flame and add the peppermint oil. Stir well. Let cool. Spoon ¼ cup of the mixture between two pieces of wax paper and flatten to make one patty. Continue to make patties until you've used up the mixture. Freeze the patties until you're ready to use them. Place one patty on the top bars between your two hive bodies in early spring.

bees right at the entrance. Females lay eggs inside the hive, eat honey and pollen, and chew through beeswax, leaving behind nasty tunnels and a cobweblike mess. They can even chew on the woodenware and spin cocoons that are difficult for the bees to remove. Often, wax moths will invade the frames after they have been removed from the hives while they are being stored in a warm place prior to the removal of the honey. Freezing your honey-filled frames for three to five days kills the moth eggs because they do not tolerate cold temperatures. Beekeeping suppliers offer a variety of treatments for wax moth.

SMALL HIVE BEETLE (SHB): These dark brown to black-colored beetles (*Aethina tumida*) are native to South Africa, but were found in Florida in 1988. Typically preferring warmer climates, they find their way into weak colonies and will eat live brood and honey. Honeybees cannot fight back or sting them through their hard shell. Female SHBs lay up to five hundred eggs at a time, overwhelming the colony, and the beetle larvae spread slimy mucous around the hive. Sometimes these vermin can cause damage similar to that of wax moths, leaving an awful smell of fermentation in a hive, an indication of infestation. Spiky larvae in and on the combs en masse cause lots of damage. Beekeeping suppliers offer a variety of treatments for SHB.

SACBROOD: Adult bees are immune to the sacbrood virus, but it affects worker and drone larvae during brood rearing in the late spring. Spotty brood patterns leave larvae as watery sacs with their heads protruding, and these young bees are left dead in the cell. This virus is thought to be transmitted by the house bees. The good news is that this virus usually goes away in late spring, but there are no treatments for it.

CHALKBROOD: Called the "mummy disease," chalkbrood is a powdery fungus, *Ascosphaera apis*, that spreads to larvae through their stomachs. The fungus was discovered in the United States in 1968 and is increased by too much moisture inside a beehive. After the fungus spreads its spores, it consumes the worker and drone larvae, leaving behind only white, pelletlike kernels ("mummies"). The house bees dispose of the kernels by removing them from the brood cells and leaving them at the front entrance of the hive. Replacing your queen, keeping the hive well ventilated and dry, and removing chalkbrood-covered frames will help clear up this problem.

EUROPEAN FOULBROOD (EFB): EFB is a stress-related bacterial disease caused by the bacterium *Melissococcus pluton* and spreads through the brood nest when young larvae eat infected food. Young larvae turn brown and die inside the cells, becoming twisted, slimy, and rubbery. This final stage is called *scale*. Once spores spread, you will see a spotty brood pattern with many concave cappings and sunken or partially capped brood. Antibiotics are a necessary treatment. Follow directions carefully when treating EFB.

AMERICAN FOULBROOD (AFB): AFB is also a bacteria, and it is caused by *Paenibacillus larvae* spores infecting young honeybee larvae through their intestines and, ultimately, their blood. This devastating disease is spread by adult bees and on infected beekeeping equipment, including hive tools and woodenware. The spores turn white larvae black, and the larvae die soon after being capped. Sticking a toothpick inside the cell, called the rope test, is a good way to test the cell. On a sunken cell, the toothpick will pull out a stringy brood that will stretch for a few inches, then pull back. If AFB becomes severe, you'll need to destroy the colony and your equipment, including the wooden hive. AFB spores have been known to survive antibiotics and can germinate after eighty years.

CHEMICALS AND YOUR HONEYBEES

Many medications have been approved for honeybee treatments and should be used only as directed on the labels to prevent them from becoming hazards to the health and livelihood of both honeybees and humans. Over the years, honeybees have become resistant to some of these approved medications, so these treatments are constantly changing. Many of the approved treatments available here in the United States might not be available outside the country or be available under the same brand names. Each country has its own set of rules for approving medications to treat honeybee pests and diseases.

Pesticides, fungicides, insecticides, and herbicides in your garden and neighboring areas can kill honeybees and affect humans in many ways we still do not completely understand. Chemicals find their way into our food through plants and farm animals and through the air we breathe. When applied to plants, they are picked up by foraging honeybees and can kill many before the bees even return to the hive. Other worker bees may carry the chemical-laced pollen back to the hive and thus taint the honey, which can then poison those bees inside the hive. Sometimes poisoned bees crawl out and die at the entrance of the hive. Most insecticides do not contain high levels of toxins that would harm honeybees, however users should exercise caution when applying these chemicals, especially around apiaries. Never treat your honeybees for disease while your honey supers on on your hive.

These environmental issues mattered to me. I learned as much as I could about natural treatments for my garden and bees. Later this led me to developing my own line of pure personal products for humans under my Red Bee brand.

Registering Your Bees

Some states require beekeepers to register their honeybees. Beekeepers are considered farmers, and honeybees are agricultural livestock, just like cows, sheep, or pigs. If there is a problem with your honeybees, such as one of the diseases mentioned above, the problem can spread to other bees and cause an epidemic. Registration allows state officials to keep track of all of the honeybees in the area. A problem can be traced back to its source and neighboring beekeepers can be made award of the problem. Should you experience a problem with your hives, your state bee inspector will come pay a visit to your apiary and inspect them.

In the United States, the registration process varies with each individual state. Usually you need to fill out a simple form, stating your name, address, how many hives you own, and their locations. The registration for my home state of Connecticut provided a space for me to actually draw the location of my apiary in relation to my house, the street, and my neighbors' homes. There is usually no charge for registering honeybees, and local bee clubs urge all beekeepers to register.

CHAPTER 10

Honeybees in Autumn: Preparing for Winter

It was a lazy August afternoon during my first season of bee-keeping. The kind of day where the sun seemed bear down on the earth and the air was still and oppressive. The lanky sprigs of goldenrod had recently expired. Their dried flower heads hung over, leaving little or no nectar for my bees. It had been drier than usual that summer, and most of the greens in my garden were on their way to a crispy brown. Late summer had officially arrived. As I approached the hive that day, I immediately noticed more than just the usual activity at the entrance. Odd, I thought, for such a lazy day. I continued forward slowly and with some caution. I had been told that in late August there might be a dearth of nectar and that the bees could become restless and hungry, cranky, and even nasty. Something referred to as *robbing* could occur. Simply put, opportunistic bees try to rob honey from another hive and carry it back to their own—a much easier way to obtain food during a drought.

Robbing behavior is considered common any time there is a lack of nectar, but rogue bees might also try to invade a strange hive as the result of sloppy beekeeping practices. On occasion, for example, I have kept the hive open too long during an inspection, thus advertising its contents to any old honeybees in the area. Simply dripping sugar solution near the hive could serve as an open invitation to unwanted guests. But honeybees are nothing if not clever and organized. For what is a queendom without a respectable army? And, indeed, guard bees are stationed at the entrance of the hive to protect the colony from enemies. As worker bees enter the hive, the guard bees wait in line at the entrance and smell each entering bee. If a bee's pheromones are not recognized, the bee will be denied entrance. The only way a rogue or drifter bee may gain entrance to new hive is if the bee approaches bearing a gift in the form of nectar or pollen.

Clearly, a strange bee had tried to enter my hive, and a brawl had broken out. As I stood at the entrance, I could actually see several bees aggressively bolting up to the entrance in an attempt to push pass the guards. In fact, the newcomers were not honeybees at all. They were yellow jackets, and their motive was obvious: to get inside the hive and indulge in the honey. I had a full-fledged robbery on my hands.

A few of my own honeybees were locked in one-on-one mortal combat with these thieves, rolling around the hive entrance and then falling to the grass, but never once breaking their combative embrace. All six legs wrapped around the foe, each bee or yellow jacket tried to sting the combatant to death. I was immensely proud of my honeybees for valiantly defending their turf. I felt like it was my duty to join in and help. So I grabbed the nearest twig and began flicking yellow jackets off the hive entrance. I further attempted to separate the brawlers. I hated the idea of my honeybees potentially giving their lives to defend their hive. But despite their courage and my feeble but well-meaning aid, the robbing continued. The next day I noticed what appeared to be crumbs at the entrance of the hive. I could not image how this pile of grit had found its way onto the entrance. When I bent down to peek inside the hive entrance, I could see where the robbers had forced their way in and uncapped some of the honeycomb. Those crumbs were actually the wax cappings that had been chewed up and left behind by the robbers, who'd most likely filled their own bellies with the stolen honey. Nature, I had discovered, could be cruel, and it made me furious.

I managed to shut down the continuation of this particular robbery for one day by closing up the entrance of the hive with a wad of grass. You are basically calling a time-out. Nobody goes into the hive, and nobody comes out. After one day, you can remove the grass. Hopefully, the robbers on the outside will have given up, and the ones trapped inside have been stung to death. If you do not catch the robbery early and it escalates, the result can be most unfortunate. Honeybees need that stored-up honey when there is a dearth of nectar and especially going into the winter months. Over the next few weeks I kept an especially watchful eye on the activity around my hive entrance.

Preparing the Hives for Overwintering

Preparing your honeybee hives in the fall for the upcoming winter is as important as taking care of them throughout the summer. A hive that has been properly prepared for a long, cold winter has a better chance of survival until the spring. As the fall arrives and foraging activity winds down, bees prepare to *cluster*.

During the winter months, the colony drops down to approximately 25,000 bees. Most noticeably, the drone population dwindles. Since the only contribution the drones make to the colony is to mate with a queen, their presence becomes rather unnecessary during the winter months, when the queen's laying is at a bare minimum. Since pollen and honey is also limited during the winter, the economy of nature takes its course and refuses to allow all the drones to remain inside the warm, cozy hive throughout the long, cold winter. As cruel as it seems, most of them are kicked out by workers and left outside the hive to starve and freeze to death. A few luckier drones are allowed to remain inside the hive during the winter for early spring mating.

The worker bees and remaining drones will form a tight cluster around their queen, which will keep her and the entire colony warm throughout the winter. Through a series of rotating movements, along with rapid wing vibrations, beginning from the inside of the cluster and working their way out, heat will be spread from the body of one tiny bee to the next. Through this process, the bees are capable of maintaining a constant temperature of 95°F inside the hive, even on the most frigid winter days. On days that are slightly warmer, the workers will crawl to the corners of frames and gently scrape open capped cells full of honey to eat. They will then carry some honey back to the cluster to feed the queen, the other worker bees, and even the few drones that have been allowed to overwinter. Some will have the unpleasant task of cleaning up after the queen and the drones. If any day happens to be 65°F or more, the worker bees will leave the hive to get a little exercise and go for quick cleansing flights. Worker bees born in the fall will overwinter inside the hive, and because their duties are light, they will live for a few months longer than their spring and summer siblings.

By the end of the summer, after the honey has been extracted from the hive, beekeepers can help their bees prepare for overwintering by feeding them with a winter sugar solution, properly ventilating the hive, and helping the colony guard against intruders. Feeding them sugar-water solution in the fall, while there is little nectar and pollen available, will help bees store up enough honey to last the winter. (A single colony of bees can consume up to two hundred pounds of stored honey over the course of a full year.) Many winter-related bee deaths are a direct result of winter starvation. In northern climates, such as where I am from, it is advisable to make sure your bees have a minimum of sixty pounds of honey stored in the frames. This honey is stored in the main two brood boxes, where it will be available to the honeybees throughout the winter. The ideal sugar solution for fall feeding is a ratio of one part sugar mixed with one part water. Bees can drink up to a gallon a day, so your feeder needs to be constantly refilled until the bees stop taking it. Feed your bees for as long as they sip the solution; they will stop when they have had enough or once they are clustering.

The fall is also the time to make any necessary repairs to the hive. Be sure to replace any damaged pieces, and make sure the cover still fits tightly, so the bees will be sheltered from rain and snow. During winter, moisture can get trapped inside your hive and create dampness. If possible, gently tilt your whole hive forward and set it at a slight angle so that rainwater or melted snow will drain out at the entrance. Some beekeepers take a wooden ice cream stick and place it between the hive bodies to create a little extra ventilation after the bees have begun to cluster. In extremely cold locations, some beekeepers swear by wrapping their hives for added insulation. Materials used for wrapping include roofing paper, cardboard, plastic tarps, or specially made colony quilts. Do not close off the entrance; bees still do need to exit and enter the hive when temperature permits. Many people ask me why I don't bring the bees inside for the winter. The reason is that it would be confusing to the bees' natural instincts, and there aren't any flowers for them to forage on inside the house.

You can also modify the entrance reducer for the winter months. Turn it to the smallest notch to allow the bees only a tiny hole to enter and exit. The smaller entrance will keep cold winds from blowing directly into the hive, keep out small creatures looking for a warm place to rest, and provide a smaller area for the guard bees to work. You also may opt to attach a *mouse guard* to the entrance. A mouse guard is a flat metal bar with holes that are large enough for only bees to crawl through. These are made specifically to keep out mice, and come spring you may be happy you used one. Mice who spend the winter inside your hive can destroy the honeycomb and eat up all the stored honey. The mouse guard can be removed again in the spring as hive population grows. If you use one during the winter, you do not need an entrance reducer. When spring arrives, you remove your mouse guard and use the entrance reducer again.

Autumn is also the time to evaluate your queen. If she is not still laying eggs into the fall to build up the colony for a long winter, it may be time to replace her to ensure your colony will be stronger and healthier the following spring. Many beekeepers routinely replace their queens each fall. If you decide to replace your queen, you first have to remove your original queen.

If you are unable to acquire a new queen, it is advisable to unite two colonies. Use the following technique: First remove the inner and outer covers of the first hive, exposing the top of the frames. Place a sheet of newspaper on top of the frames and poke a few holes in the paper. Remove the bottom board from the second hive and place it on top of the first hive. Cover the second hive with the inner cover and outer cover, making one big hive. The honeybees will gradually tear through the newspaper and unite to become one large family. As for the queens, the bees themselves will decide which of the hives' two queens will be their leader. The queens will fight it out between themselves, and the younger or stronger of the two will be crowned victorious.

REQUEENING MY HIVE

One day during the late summer of my second year of beekeeping, I was inspecting my hive as usual. By this time, I felt the confidence to undertake the chores like a seasoned beekeeper. Already I had attended many bee club meetings and received abundant support from other members. My hive had successfully overwintered its first year, and by the time spring came around, everything seemed status quo in my beeyard. But during this particular inspection, I was surprised to find a diminished number of bees in the hive and no eggs to speak of. I searched further and discovered there was no sign of the queen.

I contacted Mr. B immediately. He responded that my queen had likely died or was lost and that it would take some time before the colony would realize their queen was missing. For a while, workers would continue to fan the last of her pheromones around the hive, and the colony would carry on as usual. But once the colony realized their queen was missing, things inside the hive would change. His advice was to wait and check again tomorrow. If the queen did not appear, I would have to get a replacement queen.

The next day, with great anticipation, I opened up the hive. Still no eggs, still no queen. I began to panic and contacted Mr. B again. He gave me the name of another local beekeeper who could provide me with a new queen. Good old Eddie lived a couple of towns over. He was an elderly gentleman who had been keeping bees since what seemed like the beginning of time. He sold bees and equipment, and he also raised queens. When I called him up, Eddie told me yes, he did have queens for sale and that I could come at 3 p.m., when his shop opened. I told him that my queen was lost and that I was desperate for a new one. I said it had been two days, maybe more, since I'd seen eggs. He assured me a colony could survive a week without its queen, so mine would certainly survive until three o'clock that day.

As I waited, I consulted my beekeeping books about requeening. I read that if a hive is left queenless, a female worker will eventually take over the queen's duties and begin laying eggs. This bee is called a *laying worker*. Since worker bees are not capable of laying fertilized eggs, the laying worker's eggs would become drones. If she's allowed to continue laying eggs, it wouldn't be long before drone eggs outnumbered the worker eggs, and since drones cannot gather nectar or make honey, the hive would eventually collapse. Laying workers look just like a regular workers, making it difficult to find and remove them from the hive before requeening. One way to tell when you have a laying worker is that each cell has more than the normal single egg. A spotty brood pattern, drones in worker-sized cells, and eggs not laid at the very bottom of a cell are other indicators.

When I arrived at Eddie's place, there were already a handful of beekeepers clustered outside, waiting to pick up tools, equipment, and especially queens. Eddie's intimate shop could hold only four or five people, and the others were asked to wait in the yard. The conversation among the other beekeepers centered mostly on who'd been chasing after swarms that year and how this process sets back overall honey production. Eddie made it perfectly clear that we were not welcome to stop by unannounced and interrupt him while he was tending his own three hundred hives and making honey deliveries. He questioned each beekeeper about what was happening with his or her hives and how he could help them. I could tell he had been in this business a long time, and he seemed to know everyone there. A series of questions seemingly needed to be answered, so Eddie could diagnose each hive's behavior. "When did you last see your queen? Did you see eggs? Are there any swarm cells?" he asked.

His wife, Anita, was always at his side, helping to pack up orders for customers. Anita was a worker bee who also was a queen. Like Eddie, she was a lifelong beekeeper, and together they were the founders of the Back Yard Beekeepers Association. Once, when I had stopped by to purchase some honeycomb for candles, she showed me how to roll it up to make dinner candles.

As I waited my turn, I noticed all the honey and bee-pollen jars that were neatly displayed along the side shelves. Eddie's honey was famous around these parts, and he had been producing honey for what seemed like forever. Along with literature and educational posters was his popular book about beekeeping, appropriately titled *The Queen and I*.

As the last beekeeper left, Eddie turned to me and asked, "What can I do for you, doll?" This was his old-fashioned way of speaking to young ladies. Before I could answer, he said, "I know, I know, Mr. B sent you for a queen." Feeling a bit like Dorothy before the Wizard of Oz, and attempting to hold my own, I again told him how there were no eggs present in my hive and I did not see the queen.

Eddie offered me one of his fine young queens for a small price. She was packaged in the familiar wooden queen cage I remembered from my first package of bees. He explained that the procedure for introducing her to the hive was also the same: place the cage between two frames and let the workers and attendants eat away the candy cork to free the queen. If all went well, her new subjects would accept her, she would mate with some lazy drones, and the colony would be saved from a slow death. I listened diligently, nodding in agreement with every word.

It was after five o'clock when I arrived back home with my new queen. I had to work quickly if I wanted to introduce my new queen to the colony before dark. So I suited up, lit my smoker, and opened up the hive. I removed one frame from the top deep to make space to hang the queen cage, sugar-cork side up, closed everything up, and promised myself it would be fine. One week later I opened up the hive to find the queen was out of her cage and had already begun laying eggs. My hive was queen right again.

THE OVERWINTERING BEEKEEPER

In the course of the cold winter months, while the bees are clustering inside the hive, a beekeeper's tasks are limited. If there are more than a few consecutive warm days, you may want to feed your bees some sugar fondant or even a pollen patty. Both can be made at home and placed on the top bars of the upper deep on a warm day. Do not open your hive when the temperature is below 55°F, or you risk chilling the bees.

Generally speaking, winter is a good time to build some new frames with some fresh foundation. These can be added to the hive in the warmer days of spring to replace damaged or old comb, which becomes deep brown in color from brood rearing. I've noted that many beekeepers use this winter break to catch up on their reading. I usually pick out a few more books from our bee club library in order to learn as much as I can for the upcoming bee year.

Taking on various bee-related crafts is another great way to spend the winter. Many beekeepers clean and filter their wax for candles, make honey soap, bottle their honey, and design new labels for their honey

Sugar Fondant

Fondant is a sugar-paste feed for the bees over the winter and into the early spring. The advantage of fondant is that it will not ferment during the long winter months like the liquid sugar syrups.

½ cup water
2 cups granulated sugar
2 tablespoons light corn syrup

Line a loaf pan with wax paper. Combine water, sugar, and corn syrup in medium saucepan. Stir over medium heat until sugar is dissolved completely. Using a candy thermometer, heat to 235°F without stirring for 3 minutes. The mixture will become a soft ball, but if you overheat it, it will caramelize. Remove pot from heat and let mixture cool until it becomes thick and pasty. Pour into loaf pan. Allow to harden and slice when cool. Place a slice on the top bars of the upper deep inside hive. Save extra in plastic bag. No need to refrigerate.

Pollen Patties

Pollen patties are a good source of protein,
which is necessary for bees raising brood. I found many
recipes in my bee books, and then made up my own.
My recipe makes about one 1.5-pound patty.
Pollen patties are also good to give to bees in the
early spring since there can be limited amounts
of nectar available at this time.

DRY INGREDIENTS
1½ cups fat-free soy flour
1½ cup brewer's yeast
1 teaspoon spearmint or tea tree oil

SUGAR SYRUP
1½ cups granulated sugar
¾ cup hot water

Mix dry ingredients in one bowl and sugar syrup ingredients in a
separate bowl. Slowly add dry mix to syrup until mixture is like
stiff bread dough. Press between wax paper. On a warm day
open up your hive, remove the wax paper, and place the patty
on the top bars inside the hive, just above the cluster of bees.
You can make the patties in advance and keep them frozen to
prevent spoiling. Thaw before placing inside your hive.

jars. But my favorite activity is mixing up beeswax balms and salves.
Recipes for these projects are numerous and never hard to find.

There is something so intriguing to me about making natural
skin-care products the old-fashioned way—especially when each
product calls for only a few ingredients, most of which I have in my

kitchen already. Tired of ambiguous names of ingredients on personal-care products, I was already a big fan of sustainable products, and beekeeping led me into a new world I call "clarity"—that is, being able to read the name of each ingredient on a label and knowing exactly what it is and whether or not I want to put it on or in my body. Once I had bottled my own honey, I felt I was ready to try my hand at some other products that beekeepers have made over thousands of years. I began to see possibilities, and as I began experimenting with beeswax, I stumbled upon some brilliant combinations of natural ingredients and arrived at my own blends of beeswax-based personal products.

SOME OF YOUR BEESWAX

Pure beeswax is a nontoxic, natural wax secreted by the eight wax glands on the abdomen of the honeybees. It is used to build the incredible walls of the honeycomb. When secreted, the wax is a transparent liquid, which turns into a semisolid substance on contact with the atmosphere. It takes eight pounds of honey for a bee to produce one pound of beeswax.

Beeswax has many applications for healing. It has been used for centuries as an ingredient in natural salves and balms. It is a completely pure, natural, and earth-friendly wax. As an ingredient, beeswax is known as Cera Alba and Cera Flava and is perfect for personal-care products. When combined with oils and butters, beeswax is used as a thickening agent, emulsifier, and humectant. It can be mixed with a wide range of essential oils, herbs, and other natural ingredients to create balms that are more emollient than many commercially made products. Beeswax tends to feel better on the skin than other, refined waxes because it allows the moisturizing properties of the product to absorb and heal.

Thrifty beekeepers save every extra scrap of beeswax, because regardless of whether it is in the form of burr comb or broken pieces of foundation, all beeswax has value. Artists, crafters, sports enthusiasts, and hobbyists all use pure beeswax for a wide range of applications. I save all my scraps in a plastic bucket throughout the bee season, and while my bees rest during winter, I melt it all down and filter it for my own crafts.

The Boundless Benefits of Beeswax

Dipped candles

Molded candles

Beeswax foundation (used by beekeepers in the beehive)

Beard/moustache/ dreadlock wax

Grafting wax

Crayons

Carving objects

Lost-wax process

Batik-dyeing process

Etching glass

Ukrainian egg designing

Tack cloth

Encaustic painting

Wood filler

Polishes

Nail/screw lubricant

Brick-floor sealer

Moisturizing cream

Soap making

Fruit coating

Dental procedures

Baking-sheet coating

Embalming procedures

Cosmetics and skin-care products

Leather waterproofing

Thread and fishing-line coating

Reconstructive surgery procedures

Sealing on jams and jelly jars

Coatings for military weapons, shells, and tools

Lubricants for zippers, windows, and drawer slides

Wood sealers and finishes

CLEANING BEESWAX

It is quite gratifying to clean your own beeswax, and when you smell that warm, honey-tinged aroma, you will be glad you did. Melting and filtering your beeswax also further purifies it. To clean your wax, you'll need a large used coffee can, an old pan you don't use for cooking, a rubber band, a clean empty milk carton, and piece of cheesecloth to stretch over the top of the milk carton.

Fill up the coffee can with your wax. Place the can in the pan half full of water. Never melt wax over a direct flame or let it boil. Beeswax is highly flammable and can ignite in an instant.

Slowly bring the water to a boil till the wax melts. As the beeswax melts, it will emit that lovely aroma, the same one that you smell in the beeyard. Any debris will fall to the bottom of the can. While your wax is melting, stretch the cheesecloth over the top of the milk carton, securing it with the rubber band. When the beeswax has completely melted, carefully pour it through the cheesecloth into the milk carton. The cheesecloth will act as a filter, catching any debris that was in your beeswax. Let the freshly poured beeswax cool until you are ready to use it. Repeat with more beeswax, pouring it into another carton. The cardboard milk carton can be ripped away from the wax when it has completely cooled down and hardened. This whole process should not take more than a half hour.

MAKING BEESWAX CANDLES

Beeswax candles have always played a part in furnishing light. From early times, beeswax has played an important role in religious liturgy. The Catholic Church at one time required all religious candles to be 100 percent virgin beeswax; it regarded the wax as a symbol of purity and the Virgin Mary. It was also observed that beeswax candles burn cleanly, producing no smoke; paraffin candles emit black soot, which covered paintings and religious artifacts in the churches.

This versatile material was recognized and valued by ancient civilizations because of its excellent ability to make a slow-burning candle. Beeswax candles are sustainable, emitting negative ions that actually clean the air and invigorate the body. Beeswax candles emit light that has a light spectrum similar to that of sunlight. With the correct type of wick, beeswax candles are smokeless and dripless.

When you make pure beeswax candles from your own beeswax, there are endless types of molds to choose from. They can be found in any craft store or beekeeping-supply catalog. Don't forget to get some wicks and those little metal bases that hold the candles down as they hang from a crossbar. There are silicone molds, with amazing details, that can be peeled from the wax with little fuss.

After you clean and filter your wax, it is ready to be made into candles. First, simply remelt your clean beeswax. Using the support pin that comes with the candle mold, hang your wick in the center of the mold. Next, pour the melted wax into the candle mold, being careful not to make any air bubbles.

Ear Candling

Ear candling is a procedure that gently removes wax buildup inside your ears. Using only specifically designed candles made of pure beeswax and unbleached cotton cloth, you can do ear candling in the privacy of your own home. Many who practice this technique claim it relieves sinus pressure, improving the clarity of all their senses and giving them better balance. The process takes about ten minutes and should not be used by those who have recently had sinus surgery, have perforated eardrums, or have signs of bleeding or draining from their ears.

Instead of using molds, some beekeepers dip their wicks into cans of hot beeswax to create *hand-dipped candles*. Making this type of candle takes more patience, and it was how tapered candles were made during colonial times. As the first layer of wax dries on the wick, you dip it again until the layers build up, creating a tall dinner candle.

MAKING PERSONAL-CARE PRODUCTS

When you have all your beeswax clean and filtered, you are ready to try your hand at mixing up some lip balm and salves. The first time I made lip balm, I used a simple recipe similar to my own recipe included here. The whole process took me no more than twenty minutes, and I was amazed at how easy it was. My recipe requires very few special skills or pieces of equipment, and most of the ingredients can be found in your kitchen or local health-food shop.

When I mix up my beeswax recipes in my kitchen, I use glass bowls, stirring spoons, and other tools that are reserved *only* for making skin-care products, and I do not use those same bowls for mixing up a batch of my grandmother's favorite pasta sauce later that day. So pick up some supplies at a yard sale or secondhand shop and keep them specially for mixing your beeswax recipes. Containers for your creations can be purchased from most beekeeping-supply companies.

I was impressed and amazed at how lip balm could be made so easily and in one sitting. I gave my lemon lip balm a try and discovered it to be the creamiest product I've ever used. Knowing its ingredients were pure and came from my honeybees made it even more terrific.

Beekeepers everywhere make their own lip balms and salves using the fine beeswax of the harvest. Once you have mastered this simple recipe, you can begin to experiment with other oils and add different essential oils or fragrances to personalize your products. I could not resist any of the choices of oils on the market, and each one offered a different and healthy benefit for my finished products.

Every Beekeeper's Simple Lip Balm Recipe

INGREDIENTS:
1 ounce beeswax
5 ounces extra-virgin olive oil
5 drops Red Bee® wildflower honey
2–3 drops essential oil such as
lemon, strawberry, or peppermint

TOOLS:
Double boiler or crockpot
Stainless steel whisk or mixing spoon
Glass jars or metal tins with lids for storing final products
Scale or measuring spoons

Slowly melt the beeswax and oil in a double boiler over a low flame. Add honey; mix well with a wire whisk. Never allow your mixture to boil. Do not cover the beeswax with a lid while heating, because doing so could create condensation that will fall into your mixture and ruin it. You may also melt your ingredients inside a crockpot or a microwave. If you use a microwave oven, be extremely careful, since the wax could reach the flash point and ignite without warning, not to mention splatter all over the inside of your microwave.

When the beeswax, oil, and honey are melted and mixed well, test the texture of a sample by letting a small spoonful cool. If there is too much beeswax, the sample will be hard. Too much oil, and it will be mushy. Add more ingredients accordingly. When your sample is firm but spreadable, remove the rest of the mixture from the flame. The last touch is to add the essential oil and mix well. Using a small stainless steel or plastic spoon, pour the mixture into the storage containers. Work quickly and pour before the mixture cools down and hardens. Wipe away any drips before covering each container.

Once I feel I've perfected a recipe for a personal-care product, I offer samples to my friends and family. At first, many of them did not really understand the love that went into creating each product, or the real benefits of using pure beeswax over other waxes. It was a few years before my natural products gained a following. I found most of my first loyal customers at farmers' markets, places where people come to purchase their food and other products firsthand from the producers. There I could let shoppers test my honey and beeswax skin-care products, and I could get their reactions firsthand. After customers tried my products, they returned to report how wonderful and pure the products felt on their skin. I was spreading the good word about honeybees

The Buzz on Beeswax

1. Beeswax is solid at room temperature and melts at around 145 to 147°F, the highest melting point of all waxes. Beeswax candles burn longer than other types of candles.
2. You know you have pure beeswax when a whitish, powdery deposit appears on the surface of your wax after it is stored for some time. This powder is called *bloom,* and it can be removed simply by rubbing the wax with a soft cloth or gently applying heat with a hair dryer.
3. The expression "It's none of your beeswax" is a common phrase, often said to young children, that means "It's none of your business." It may have stemmed from a time when only the upper class could read and write. Before envelopes were used for written correspondence, letters were folded and sealed with a drop of beeswax and then hand stamped with a ring depicting the family's seal. This sealing method insured privacy while the letter was in transit.

and educating folks about using pure products. The more customers learned about honeybees and my natural creations, the more they came back to purchase my products. Before I knew it, new customers were stopping by my market table each week and specifically asking for one of my beeswax salves. The general public began to understand what I have long believed: what goes on your skin is as important as what you put into your body. Every day I spoke with customers who had developed allergies and health and skin conditions related to their diet and the environment. Every year more people seem to be seeking out honey and bee products to help them alleviate these ailments. Through my research in China, the UK, and Italy, I was able to develop the purest skin-care products available in this country, using vegetable-based oils and butters combined with beeswax and honey.

Between taking care of my hive and mixing up my beeswax salves, I was devoting more and more time to my honeybees and paying less attention to my day job. Also, the nature of the giftware-importing business I worked for changed rapidly; the company eventually decided to close down its wholesale and design departments and focus on retail. Luckily, this change gave me an opportunity to take the plunge into beekeeping and making and marketing my products full-time. Why not? I had been developing and marketing products successfully for other companies, and it was time for me to take on the role of the queen bee.

CHAPTER 11

Apitherapy:
How the
Honeybee Heals

Before completely leaving my job with the giftware company, I made one last trip to China. Spending eighteen long hours on an airplane, breathing in all that stagnant air, always seemed to bring on sniffles and a cold, so I wasn't surprised when, that evening, I felt a sore throat coming on. In China they do not have pharmacies as we know them, but there are full-service apothecaries almost every few blocks, and there is always a qualified herbalist or doctor on staff to answer any questions and recommend herbal remedies. These pharmacies are similar to our twenty-four-hour emergency-care clinics, but not half as intimidating or expensive. It goes without saying, you need to be accompanied by a bilingual speaker to translate your needs to the medical staff. Being able to obtain teas and herbal remedies by a mini private consultation had always appealed to me, and since I had been keeping bees for a while, I was particularly on the lookout for any bee remedies. Propolis extracts, royal-jelly tablets, and bee-pollen jars were just a few of the antidotes I found at the Chinese apothecary. I wondered why we did not have such products so readily available back home in the United States.

I explained my throat and head discomfort to my business associate Mr. Wang, who translated to the herbalist. He recommended I take some loquat-honey syrup for my throat and cough, and a tiny bottle of clear liquid essential oils called Bai Wan You for headache relief. Without hesitation, I opened up the tiny bottle of Bai Wan You (white flower oil) for a sniff. Clearly, it was peppermint and eucalyptus oils to rub on your sinuses and temples. It felt really good, soothing, and wholesome knowing exactly what was in the bottle. Upon arriving at our hotel, I immediately took a swig of the loquat-honey syrup and was surprised at how mild and pleasant it tasted. It was the most delicious cough syrup I had ever tasted. The ingredients list stated it was simply a mixture of honey and loquat, a Chinese plum. It was soothing because of the honey and pleasant because there wasn't any alcohol in it, as there is in cough medicines back home in the United States. This got me wondering why U.S. cough medicines had alcohol in them at all. In China, I found honey used in cough medicines, herbal teas, candies, and even facial creams. I began to notice honey was more important than sugar as a sweetener, and I liked that.

During this working trip, I was wearing what I call my honey eyes, meaning I began to see honey and bees everywhere. Unbeknownst to me at the time, honeybees have a rich history in ancient China. I purchased as many honey items and other bee products as I could find to take home with me. The acquisition and understanding of such products plainly became an obsession, and I surrendered to it. The Chinese taught me many things, especially about natural, honey-based remedies.

APITHERAPY

Apis means bee in Latin, and *apitherapy* is the ancient practice of using bee venom and other products of the honeybees, including honey, pollen, royal jelly, and propolis, along with essential oils, to treat a variety of conditions and to maintain health and harmony within the body. My enthusiasm for honey and its supernatural healing properties led me to the American Apitherapy Society (AAS), a nonprofit membership corporation that promotes and educates people about the use of honeybee products to maintain and improve health. I found this ancient wisdom riveting, so I signed up and attended one of the society's yearly conferences. Theo Cherbuliez, a native of Switzerland and a psychiatrist and physician who served as president of the AAS for more than a decade, led the conference.

The conference is a series of workshops and training called the Charles Mraz Apitherapy Course (CMAC). Charles Mraz was a beekeeper and author who founded the AAS in the 1930s and served as its executive director. During his lifetime, he used bee stings to treat people with arthritis pain, multiple sclerosis, and other autoimmune diseases. His book *Health and the Honeybee* was published in 1995 and includes his personal journey as a pioneering healer along with case studies. Mraz was once voted one of the five most distinguished beekeepers in the United States, and he traveled around the world, educating people about the amazing honeybee.

Over the course of the three-day conference, I listened to speakers who had come from all over the globe to offer their insights on apitherapy. They were beekeepers, doctors, educators, practicing apitherapists, acupuncturists, as well as people recovering from illnesses from which only the honeybee and her products offered relief. All of these interesting and knowledgeable people had gathered to talk about honeybees and their products for healing. There was a vast amount of information to absorb.

BEE VENOM THERAPY

Bee venom therapy has been used throughout history and has helped cure many ailments. Hippocrates, the father of medicine, used honeybees and their venom to treat patients more than two thousand years ago. The use of live bees to literally sting certain pressure points, similar to the technique of acupuncture, has its roots in ancient Chinese medicine. Acupuncture balances the flow of energy and promotes natural healing through the insertion of extremely fine needles into the body at precise points. The placement of these needles activates the meridians, or energy channels, to move and balance energy within the body, which in turn relieves the symptoms of many conditions and promotes general health and well-being. Bee venom is administered through a systematic protocol of stings and following the same strategy of pressure points used in acupuncture. More important, it follows neurological trigger points. It is thought that bee venom jump-starts the immune system by improving blood circulation and by stimulating the body's natural ability to produce cortisone, which is produced by the adrenal gland.

Bee venom is said to be very effective for aiding multiple sclerosis and rheumatoid arthritis. It has also been applied to a range of other conditions, including sciatica, migraines, and lupus. It has been reported that bee venom has been used to kill cancer cells when all other treatments fail. A trained apitherapist will carefully profile a patient before applying any bee venom. Bee venom is not a substitute for a good diet and exercise. There is a long list of medications, including beta blockers, which suppress the immune system, anti-inflammatory drugs, as well as alcohol, that should be cleansed from the body before bee-venom treatment begins. Although this technique has not been approved by the traditional medical community here in the United States, many countries around the world, including some in Europe, recognize apitherapy as a medical treatment and routinely use it as a way to heal.

PASS THE BEE POLLEN—PLEASE!

Bee pollen comprises those brightly colored morsels that I had observed my honeybees carrying into the hive. Pollen is the male sex cells of a flowering plant, and honeybees gather it in the pollen baskets on their hind legs. They mix it with nectar and their own enzymes (invertase) and bring it back to the hive as a tiny granule, now called bee pollen.

Bee pollen actually comes in all sorts of colors, ranging from yellow ochre to burnt sienna, depending upon the type of flower it is gathered from. Bee pollen tastes like edible flowers. When eaten fresh, it is slightly sweet with notes of honey and beeswax. Bee pollen is said to be nature's most complete food, containing every nutrient needed to sustain life. This complete source of protein contains many of the basic elements in the human body, including twenty-two amino acids, twenty-five minerals, fifty-nine trace elements, eleven enzymes or coenzymes, fatty acids, carbohydrates, and protein. Among the eighteen vitamins it contains are B-complex vitamins and vitamins C, D, and E. It is said that man can survive a full three months on just two tablespoons of bee pollen a day. Since the time of the first Olympics, athletes have been taking bee pollen for sustained energy and vitality. Bee pollen is also said to decrease symptoms related to hay fever and seasonal allergies. It is recommended that new users of pure bee pollen start with smaller amounts to make sure there are no reactions, then the intake can be slowly increased. Bee pollen should be crushed for easier absorption before taking orally.

Bee pollen is said to regulate blood pressure and lower cholesterol. Crude, raw honey also contains traces of bee pollen that are beneficial to humans, and some veterinarians recommend pollen for certain animals. Beekeepers collect pollen by placing a wooden, drawerlike piece of equipment called a *pollen trap* at the entrance of the hive. As the bees return to the hive with pollen on their hind legs, they are encouraged to crawl through a narrow tunnel of the pollen trap, which pops the pollen off their legs. Losing the pollen must be confusing to the bee, and beekeepers should not attach the pollen trap too often, as bees need pollen to feed themselves and their brood.

PROPOLIS: THE PERFECT ANTI-EVERYTHING.

It is said that the third wise man, who carried myrrh to the birth of the baby Jesus, actually carried propolis. Famed Italian violin maker Antonio Stradivari polished his violins with beeswax and propolis. Aristotle is noted for giving propolis its name; *pro* means "before" and *polis* means "city," so *propolis* translates literally as "before the city," or loosely as "in its defense." The ancient Greeks observed how honeybees used propolis as a defense for the bee "city," or hive; in fact, honeybees practically weld their hive joints together with propolis before the oncoming winters to fortify the hive against harmful bacteria. Today personal-care items, like topical salves, throat sprays, tablets, and toothpaste found in most health-food stores, tout the benefits of propolis. It has been proven to have antibiotic, antibacterial, antifungal, antioxidant, and antiviral properties, all wrapped up in one ultrasticky ball. Propolis contains all the essential minerals, plus iron, calcium, aluminum, and manganese; it is rich in B vitamins and all of the other vitamins known to man except vitamin K. It is known to stimulate the body's immune system and is also found in trace amounts in raw honey, especially honeycomb.

In its raw, natural state, propolis is highly aromatic and bitter tasting, gooey like taffy, and can be kneaded into balls when it is warm. It cracks like peanut brittle when cold. Propolis can be softened or diluted with alcohol, and poured into bottles and sold as a tincture. Colors of propolis range from deep golden yellows to reddish brown to almost black. As with pollen and honey, the plant it comes from determines its color and composition. It is a well-known fact that the earliest beekeepers were the ancient Egyptians, who embalmed their mummies with propolis, and for this reason many of those mummies have survived for thousands of years. Propolis is widely used topically in many eastern and European countries to promote new cell and tissue growth in open wounds. Rosemary propolis and red propolis from

Brazil and Cuba are made by Africanized bees, who gather resin from two rainforest plants, green Alecrin and Clusia, respectively. These two types of propolis have recently gotten some attention.

One other interesting fact about propolis is that if a wayward mouse enters a beehive, it often has no defense against the worker bees, which sting it to death. The bees then completely cover the mouse in propolis before the rotting carcass spreads its odor or bacteria throughout the hive.

LET THEM EAT ROYAL JELLY

Yet another magical product made by honeybees, royal jelly is the food of the queen bee. I had seen royal jelly in plenty of ginseng drinks, facial creams, and capsules in natural-food shops, but like most people, I really had no idea what it was or where it came from. This white, creamy liquid is produced by nurse bees, who secrete it from glands in their heads and feed it to all bee larvae. The queen bee, however, will continue to eat royal jelly her whole life. The ongoing consumption of royal jelly is one of the reasons queen bees are twice the size of worker bees, lay about two thousand eggs a day, and live three to five years longer than workers. To gather one ounce of royal jelly, you would have to visit 120 queen cells and gently harvest it. This explains the rarity and cost of this magic elixir.

Royal jelly is a very rich source of both essential and non-essential amino acids; vitamins A, C, D, and E; trace minerals and calcium, copper, iron, potassium, phosphorus, silicon, and sulfur; essential fatty acids; and sugars. These are some of the necessary ingredients humans need to maintain a healthy immune system. In traditional Chinese medicine, it is said that royal jelly stimulates the reproductive system and promotes good health and well-being. Royal jelly also contains collagen and lecithin, which benefit the skin and make royal jelly a popular ingredient in skin-care products. It also contains cholesterol-lowering components, such as all the B vitamins, pantothenic acid, phytosterols, enzymes, and acetylcholine, a neuronal transmitter. In addition, royal jelly is said to aid in mood disorders, relieve insomnia, repair nerve cells, strengthen liver functions, and soothe digestive disorders. A potent antibacterial protein called *royalisin* has also been found in royal jelly.

One thing I can tell you for sure is that royal jelly tastes terrible in its authentic, fresh frozen state yet it should be taken orally when it is as fresh as possible.

HEALTH BENEFITS OF HONEY

Raw honey is a living food with unadulterated health benefits. The pH of honey is commonly between 3.4 and 6.1; the average is 3.9. This relatively acidic pH level and honey's low moisture-content prevent the growth of many bacteria responsible for infection. They also keep honey from spoiling. Honey is a predigested sweetener that is easily assimilated by the body, and it is an excellent energizer that can be used to enhance athletic performance and relieve low blood sugar. Honey supplies two stages of energy. The glucose in honey is absorbed by the body quickly and gives an immediate energy boost. The fructose is absorbed more slowly, providing sustained energy. Raw honey still in the comb contains trace amount of bee pollen and propolis; it is excellent for your immune system. Raw honey has been proven to kill *E. coli*, staph, *P. aeruinosa*, and *H. pylori*, which causes many stomach ulcers. The enzyme glucose oxidase in honey makes honey a natural preservative and can create hydrogen peroxide to form an antimicrobial barrier. Honey has antiseptic, antibiotic, antifungal, and antibacterial properties, and it is a source of antioxidants.

The healing properties of honey have earned it a reputation as one the purest and most natural remedies. Honey has been used for centuries to treat a wide range of medical problems, including wounds, burns, and scrapes. When honey is used for wounds, inflammation, swelling, and pain are quickly reduced, and healing occurs rapidly with minimal scar tissue.

Honey local to your area contains minute particles of pollen from the local flora. By ingesting pure, crude honey that is produced by honeybees in your geographic locale, your body begins the process of desensitizing itself, and you build up a natural immunity to dust, mold, and pollen in the air.

The following are some things that I've learned throughout the years about honey from beekeepers here and abroad. Many of these honey remedies have helped me personally, and I often turn to them before taking any over-the-counter medications.

HONEY AS A COLD REMEDY: Honey is a natural remedy for the symptoms of nasal congestion, sore throats, and flu. Whenever I have a sore throat, I take a spoonful of honey, or I gargle with a mixture of two tablespoons of honey, four tablespoons of cider vinegar, and a pinch of salt. Honey coats your throat, instantly making it feel better. Adding a little eucalyptus oil or fresh ginger will help to ease congestion. Opera singers use pure honey to soothe their throats before performances.

HONEY FOR SINUS PRESSURE AND ALLERGIES: Honeycomb or raw liquid honey, when ingested, can alleviate sinus pressure within minutes. Mixing honey with some fine-grained salt in warm water and pouring through the sinuses using a neti pot is an ancient Ayurvedic technique popular in India and South Asia—and it really works. *Neti*, which in Sanskrit literally means "nasal cleansing with water," is the practice of using a neti pot to irrigate the sinuses. Why sniffle and sneeze when you can clear it out?

HONEY AS AN OINTMENT: Honey can help to keep external wounds, such as cuts and minor burns, clean and free from infection and can minimize scarring. Because of the hygroscopic qualities of honey, it absorbs water and causes the skin to hold moisture. It also acts as a mild antiseptic, so honey can help to prevent the growth of bacteria. In July 2008, the U.S. Food and Drug Administration gave clearance to a manufacturer of products of wound and skin care to sell active manuka honey for medical use.

Manuka honey, found only in New Zealand, is made from the flowers of the tea tree bush and is known for its high antibacterial properties. It has been shown to inhibit the growth of the bacteria responsible for stomach ulcers and dyspepsia, including *Heliobacter pylori* and other gastrointestinal bacterias, like *E. coli* and *Streptococcus faecalis*. This marks the first time the FDA has approved a honey or honey-based product as a medical treatment specifically for first- and second-degree burns and traumatic and surgical wounds.

HONEY FOR DIGESTION: It was the Romans who first discovered honey's beneficial effects on digestive disorders. They would prescribe honey as a mild laxative. Honey has also been used as a treatment for upset stomachs, gas, indigestion, diarrhea, stomach ulcers, and constipation. Honey is believed to help destroy certain bacteria in the gut by acting as a preserving agent.

HONEY AS A CLEANSER: A cup of hot water mixed with apple cider vinegar, freshly squeezed lemon juice, and a teaspoon of honey will act as a liver cleanser while boosting your energy levels.

HONEY FOR QUICK ENERGY: Honey is easily absorbed by the body and is source of natural, unrefined sugars and carbohydrates, which provide both an instant energy boost and long-lasting effects. For this reason, many athletes eat honey during training sessions.

CHAPTER 12

All About Honey

The United States alone has more than three hundred different varieties of honey—more than any other single country. Honey is produced in every one of our fifty states; North Dakota, California, Florida, South Dakota, Minnesota, Montana, and Texas are the top producers. Our vast range of flora and fauna and our variety of climates and weather contribute to the many varietals of honey available in this country.

One can travel to practically any part of the world and find beekeepers keeping bees and harvesting honeys unique to their region. There are thousands of honey varietals found around the world, each with its own distinctive characteristics and flavors. Besides having different flavors and characteristics, honey comes in a wide variety of forms and styles, such as liquid, creamed, and comb or chunk honey— each a culinary delight of its own merit.

The United States imports as much honey as it produces domestically. Forty percent of U.S. honey is used at the table, and the other 60 percent is used in food manufacturing. The average American eats approximately 1.3 pounds each year, and this number is growing. In general, honey is experiencing a renaissance, and Americans are consuming and appreciating honey more than ever. Restaurants and gourmet food shops have awoken to the call, and honey is now following the trend of artisanal wine, cheese, coffee, tea, and chocolate. And beekeepers are the stars. Honey is more common than you think, and it is easy to find unusual types.

It is sustainable, natural, and good for you, and in your own area, you can often find honey made by local bees and harvested by local beekeepers.

What Exactly Is Honey?

Most of us can't resist a little something sweet from time to time, and some of us have a serious sweet tooth. So it's no wonder many of the foods we eat are presweetened with white sugar, which is now one of the top three ingredients in most packaged foods. Most of us don't realize how much sugar we're are actually consuming every day because so much of it is hidden in other foods, and its presence is buried in long ingredient lists. A twelve-ounce can of a typical soft drink, for example, contains about nine teaspoons of refined sugar. White sugar is overprocessed and composed of sucrose, a white granular crystal refined from cane or beet juice by stripping away all its vitamins, minerals, protein, fiber, and water. It contains very little nutritional value and promotes the growth of bacteria that cause disease and deplete your body of necessary nutrients. Organic and brown sugars have no more nutritional value than white granulated sugar. I am particularly sensitive to white sugar. It tends to give me a headache, causes me to feel fatigued, and seems to trigger mood swings. Honey does not have the same effect on my body as plain white sugar, and substituting honey for sugar has helped me eliminate processed sugars from my diet almost entirely. Unfortunately, most people assume that honey is just another type of white sugar, and some folks think all honey is homogeneous. I am certain they think this way because the only honey they've ever tasted is from a squeezable bear or the plastic honey packets found in diners.

The United States National Honey Board defines honey as a pure product that does not allow for the addition of any other substance, such as water or other sweeteners. Pure honey is *not* the same as cane or beet sugar. It is not refined. Honey is a natural sweetener made from the nectar of flowers. It is made up of carbohydrates and water, yet it also contains amino acids, vitamins, and minerals, like calcium, copper, iron, magnesium, manganese, niacin, pantothenic acid, phosphorus, potassium, riboflavin, and zinc. Honey is sweet; it does contain fructose, sucrose, and glucose, as well as small quantities of other sugars. Glucose and fructose are simple sugars converted from sucrose by the bees. Fructose is also found in fruit and converts to energy more efficiently than white sugar. Honey will not make your blood sugar rise or fall as rapidly as processed sugar. Pure, wild honey is considered a living, raw food, which means it is uncooked, unheated, and commonly, unpasteurized. So, if you have a sweet tooth, you should give honey a try.

RAW HONEY VS. HEATED HONEYS

The freshest honey, in my opinion, is honey taken straight from the honeycomb. Most beekeepers or honey connoisseurs would agree. The honey inside honeycomb is raw honey. Untouched by human hands, it is still in its original state, exactly as the honeybees made it. However, is not uncommon to find jars of liquid honey that is called raw even though the honey has been partially extracted from the beeswax comb. Liquid honey that is designated as "raw" often contains cappings and even bee parts.

Many larger honey packers or importers will filter honey until it is completely clear, so there is less chance of crystallization while it is on the shelf at your grocery store. These manufacturers know that most

Americans will not buy honey that is granulated, because granulated honey is commonly thought of as outdated or spoiled. But this is actually not true.

Honey that is heat-treated to delay crystallization cannot be considered to be truly raw. Commercial producers often raise the temperature of the honey to 170°F (77°C) for two minutes, then rapidly cool it to 130°F (54°C). Other heat treatments include heating honey to 140°F (60°C) for thirty minutes, or 160°F (71°C) for one minute, or some straight-line gradient between those two temperatures. Any more heating is considered pasteurization. I believe that heating honey to a temperature higher than what naturally occurs inside the beehive, 90° to 95°F (32° to 35°C), compromises the quality of the honey, destroying valuable enzymes and flavors. A burned flavor can occur when a honey has been overheated to kill the yeast in it or simply to liquefy it. Beekeepers who use high temperatures during the extraction process can scorch the honey and damage the flavor. Burned honey tastes similar to a caramel candy. However, darker-colored honeys, such as buckwheat, have color and flavor profiles that give it a naturally burned flavor.

CRYSTALLIZED OR GRANULATED HONEY

We've all have found a jar of cloudy, coarse granular-looking honey hiding in the back of the cupboard and thought it might be a good time to toss it out. Honey sometimes takes on a semisolid state known as *crystallized* or *granulated* honey, but it hasn't gone bad. Although most honeys will crystallize in time, the crystallization process can be avoided or delayed with proper storage. At room temperature, crystallization can begin within weeks or months. Warm temperatures of 70 to 81°F (21 to 27°C) discourage crystallization. Very warm temperatures that are over 81°F (27°C) reverse crystallization but also degrade the honey by removing valuable enzymes.

Honey is a supersaturated sugar solution out of which the glucose will crystallize in time. This natural phenomenon occurs when glucose, one of three main sugars in honey, loses water and forms a solid crystal. This supersaturated state occurs because there is a more than 70 percent sugar and less than 20 percent water content. Crystallization can also occur when tiny particles of dust, pollen, or even air bubbles act as seeds inside the honey. The honey solution will naturally change to the more stable solid state of crystallization. Tupelo, sage, and sourwood honeys are valued because they almost never crystallize.

CREAMED HONEY

Beekeepers can control the crystallization process to create a mouthwatering, smooth product called *creamed, spun,* or *whipped honey,* which is delightful to spread on toast. This type of honey has a pleasant texture and very delicate crystals in it, unlike that coarsely granulated honey in the back of your cabinet. The creaming process starts with pure liquid honey extracted from the comb. First, it is strained through a stainless steel mesh strainer to remove any debris that can cause coarse granulation. Then the honey is heated to 140°F (60°C) for one minute. To prevent overheating, the honey must be stirred constantly. This dissolves any crystals that may be present. Next, a small portion of already finely textured creamed honey is added to the liquid honey. This step is called *seeding* the honey so the same fine crystals will grow inside the newly heated honey. After being mixed well, the heated honey is placed in a cold room of about 57°F (14°C). Within two weeks the honey should completely crystallize to a luscious, creamy texture. If it is too hard to spread, it should be moved and stored at room temperature until it becomes spreadable. Crystallized honey stored at high temperatures will return to its liquid state, because some of the crystals will dissolve and not form again. Honey appears lighter in color after crystallization.

Fermentation

If you've ever opened up a jar of honey only to find that it smells a bit like vinegar, it has probably fermented. All honey contains yeast cells. Fermentation occurs when the yeast cells divide and multiply in honey that contains more than 19 percent water. This can occur when a beekeeper removes the honey from the beehive before the bees have completely removed all the excess water from the honey. Or, if you drip some water into a jar of honey by mistake, the next time you open it, there is a chance it will have fermented. Honey also absorbs moisture in humid conditions, which can cause fermentation to begin. Moisture is one of the two archenemies of honey, the other being too much heat.

Organic Honey

Organic has become a household word, and many people seek out and prefer organic foods and products. Presently, the U.S. Department of Agriculture regulates the labeling of honeys as organic. Among other requirements, it states that the management of honeybees and production of organic honey cannot use chemicals such as pesticides, insecticides, and herbicides. Organic honey producers must determine that no chemicals are being used where their bees naturally forage. But since honeybees may travel up to two miles from their hive to gather nectar and pollen, how can we know for sure that they're not coming in contact with chemically treated plants? Feeding bees sugar-water supplements in spring and fall, when there is no nectar flow and pollen naturally available, is also prohibited in organic beekeeping unless the sugar is organic. Beekeepers often treat their bees with FDA-approved medications because honeybees are living creatures and they occasionally play host to pest or diseases; the use of these mainstream treatments is prohibited in order to qualify for an organic certification of your honey. Finally, beehives must be painted with nontoxic paints.

Besides adhering to these strict requirements, beekeepers need to file a potentially costly formal application to certify their honeybee hives as organic. They are also required to keep records and be inspected each year. Honey that is properly labeled organic could be traced to the honeybee hive it was harvested from. As a result of all of these rigid standards, truly organic honey is nearly impossible to find in the United States.

FAIR TRADE HONEY

Fair trade is a movement that promotes paying farmers fair prices for their products, especially when those products are exported from developing countries to more developed countries. This policy makes fair-trade honey a wise and responsible choice. The money makes a huge difference in the lives of the farmers and their families, especially in poorer countries. Criteria like healthy working conditions, sustainable farming, and fair-trade terms are strictly regulated. Fair-trade products are not necessarily organic, but higher prices are paid to those farmers who integrate sustainable techniques, like recycling and composting. Fair-trade organizations forbid the use of genetically modified (GM) seeds, and the organizations enforce this regulation as best they can.

KOSHER HONEY

The Torah refers to the ancient country of Israel as "a land flowing with milk and honey." Honey is a quintessential part of the Jewish New Year, Rosh Hashana. Apples and challah bread are traditionally spread with honey. Although honeybees are not kosher animals, honey is considered kosher, because even though bees bring honey into their bodies, honey is not a product of their bodies; in other words, honey is stored in bees' bodies, but not produced there. For honey to be considered kosher, it needs to follow the same stringent preparation methods as all kosher foods.

STYLES OF HONEY

Besides the ubiquitous liquid form, honey comes in other forms and consistencies. By understanding each different style of honey available, you will know what you are purchasing. Here are the most common styles of honey found in the marketplace.

VARIETALS/MONOFLORA/SINGLE-FLORAL honeys are made from a single species of flower. Because each variety of flower has a particular type of nectar, the honey made from the nectar of that flower has a distinctive flavor. For example, honeybees that collect nectar exclusively from an orange grove will make orange-blossom honey. From a forest of sourwood trees, they will make sourwood honey. For a company to label a honey as a varietal, or single-source, honey, the product has to consist of 51 percent of that specific type of nectar. The remainder can be made from any other type of nectar from the same fields from which the original nectar was collected. Honeybees can be managed to forage in specific fields to gather nectar from a single source. Beekeepers know when certain flowers are blooming in their area and that their bees have a short window of time to work that particular nectar. Once the nectar flow has expired and the flowers have dropped, beekeepers quickly remove the honey supers from their hives to insure a single-floral-source honey.

BLENDED HONEY is what we find most commonly in our local grocery stores. It is honey from different floral sources blended together to create a pleasing flavor. Larger honey packers and distributors create a consistent flavor profile for the masses by using the more commonly available and less distinctly flavored honeys. Some of these honeys are imported from other countries. The final product tends to be a clear and overly sweet honey without any distinctive flavor characteristic. Unfortunately, many people who say they dislike honey feel this way because blended honey is the only type they have ever tasted. I have converted many honey loathers into honey lovers by offering them a taste of my own single-floral honey. In most cases they were surprised by the difference and are now on their way to becoming honey fanatics.

LIQUID HONEY is the honey most people in the United States are familiar with and pour into their tea. Easy and convenient to use, liquid honey is usually sold in the popular squeeze bear or those familiar oval jars beekeepers call *queenline*-style jars. Liquid honey can be single floral or blended. All of the beeswax has been filtered out during the extraction process. Larger honey packers may also heat or pasteurize their honey, eliminating many of its vitamins, minerals, and other valuable properties. Liquid honey purchased from your local beekeeper may vary in color and, if it has not been filtered excessively, it will appear cloudy, retaining particles of pollen and propolis. It is not unheard of finding a bee part or two in unfiltered, raw honey.

COMB HONEY, HONEYCOMB, OR SECTION HONEY is what I call the jewel of the beehive. In *The Origin of Species*, Charles Darwin describes honeycomb as a masterpiece of engineering that is "absolutely perfect in economizing labor and wax." Architecturally, the hexagon is the most efficient shape, using the least amount of beeswax.

Honey in the comb is uniquely delicate and light because it is still inside the wax where the bees stored it. When you spread honeycomb on a slice of bread, the honey oozes out of the tiny wax cells, exposing it for the first time to the air. This uncompromised freshness is why I call honeycomb the purest, rawest form of honey. As it oozes out of the cells, the honey can be spooned into your mouth for a sweet explosion. And, yes, you *can* eat the wax.

A perfect honeycomb specimen has no uncapped cells, dry holes, drips (called *weepings*), or damage from bruising. It should appear smooth and consistent in color. Honeycomb can be round or square. Each section usually weighs not less than twelve ounces. All liquid honey is spectacular, but once it is filtered from the wax and poured into a jar, it will never have the same delicate taste as fresh honeycomb.

CHUNK HONEY is a chunk or piece of honeycomb floating inside a jar of liquid honey. If you measure a frame from a honey shallow, you'll see it is possible to cut out two pieces of honeycomb that are four inches by four inches, leaving a narrow piece left over. This extra piece is what is reserved for chunk honey, and therefore no part of the honey frame is wasted. That piece should be placed inside the jar perfectly vertical with the beeswax cells pointing up from the center foundation piece. Consumers can either pour the liquid honey out from around the comb or scoop out a chunk of the comb itself. Chunk honey is like having the best of both worlds. This style of chunk honey is one of my favorites and would be my honey of choice, along with a piece of toast, if I were stranded on a desert island.

DRIED OR DRY HONEY is honey that has been dehydrated with drying aids and then processed. It is sold as powder, flakes, granules, or crystals and can be light or dark in color. Dry honey has added products to stabilize it and is not considered pure honey at all. There could be sugars, corn syrups, processing aids, bulking agents, or anticaking agents added, making the total content of honey only 50 to 70 percent. Unlike pure honey, dry honey has a limited shelf life, and probably the only benefit is the convenience of the powder and the fact that it's not as messy or drippy as liquid honey. If you are a purist, like myself, and prefer the real thing, powdered honey is not for you.

CRYSTALLIZED OR CREAMED HONEY is a spreadable honey with a lovely granular texture that dissolves on the tongue. It is intentionally crystallized to give it the unique quality of being smooth and rough at the very same time. Creamed honey is a favorite in Europe and many countries and makes a sinful treat when spread on toast and sprinkled with cinnamon. Crystallized honey appears creamy and almost opaque and white in color. The process for forcing honey to granulate (see earlier in this chapter) was developed and patented by Professor Elton J. Dyce in 1935.

FRAMES OF HONEY are exactly that: full frames of honey taken right out of the beehive. The honey is still in the cells, and wax is still attached to the frame. Untouched by human hands, it is a rare and special item that will probably cost you a pretty penny. If you are inclined to enjoy your honey directly from the beehive, still in its original bee frame, ask your local beekeeper. I have had honey lovers and gourmet food chefs purchase full frames of honey for eating or just as an extravagant display. A full frame of honey is a stunning sight. It is considered a delicacy in some Middle Eastern countries.

INFUSED HONEY is honey that has flavors steeped or infused into it to enhance its natural flavor. Most commonly added flavorings are fruit flavors, herbs, spices, or essential oils. Be sure to read the labels carefully to see if the honey you are purchasing is the authentic varietal or an enhanced product with additives.

CHAPTER 13

Honey Sommelier: The Tasting of Honey

W hat comes to mind when you think of the Amalfi coast? Sun-drenched beaches? Glorious gardens? Or apiaries, perhaps?

My appetite for adventure and passion for beekeeping began to take me on trips that afforded me the chance to explore the local honey culture of places around the world. Not long after becoming a beekeeper I traveled to Italy, where apiculture is a well-established agricultural activity and a way of life. After a short car ride over the hills north of the Amalfi coast, I arrived at the charming Bottega della Api, or the Little Bee Shop, in Cava di Tirreni, Italy. Beekeepers Giovanni and Francesco run a small organic apiary and small farm that includes more than three hundred hives. Neither of them speaks a word of English.

I had e-mailed ahead of time in order to arrange for a honey tasting, but had not received a response, so I arrived unannounced with the help of directions from a stranger I met at a nearby café. I found Giovanni in the driveway, and I introduced myself in his native tongue. When I told him I was a beekeeper, Giovanni broke into a warm smile. I explained to him that I had e-mailed Francisco to say I would like to schedule a tasting and that perhaps he was expecting me. I was invited to wait inside the honey shop while Giovanni called Francesco on his mobile phone. Inside the shop were humble displays of aromatic beeswax candles in all sorts of shapes and sizes, dark bottles of propolis, and of course, a large selection of bottles and clay pots filled with honey from the apiary. Each container was adorned with a simple grass tie and a handwritten tag. The clay pots were all hand painted with pictures depicting the Amalfi coastline. I was familiar with the chestnut, eucalyptus, and acacia varietals that were on display, but had not heard of *rosmarino* and *corbezzolo*. "What could these be?" I wondered.

Francesco entered the shop with a warm *buona sera*. I apologized for not calling ahead, and in a typical Mediterranean fashion, he welcomed me despite my unannounced arrival. He escorted me into the tasting room to sample the many varieties of honey they produced. The tasting room was a modest area inside a larger barnlike structure that housed all types of industrial-looking extracting equipment, honey shallows, and boxes full of freshly bottled honey. A stainless steel circular tank was full of dark, luscious honey, and a spigot was turning out the freshly strained goods. The aroma was intoxicating, and Francesco told me what I smelled was the *castagna*, or chestnut honey. Francesco passed me a ceramic jar of brightly colored, long-handled tasting spoons—the same sort that are used for Italian gelato. We began with the honey lightest in color: *acacia*. Acacia was a common tree in Italy, and although the honey color was almost absolutely clear with only a slight yellowish tinge, the flavor was unexpectedly strong. It was very clean with a hint of grassiness, and there were also wonderfully unexpected notes of pineapple and butterscotch.

Next, I really wanted to taste the *corbezzolo*, which turned out to be honey from the Mediterranean arbutus plant, similar to our strawberry bush. It was heavy in flavor, bursting with fresh berries and quite perfumy and bitter on my tongue. I sensed sweet berry notes in the aftertaste. It would have been delicious, I thought, drizzled over some chocolate gelato. The next selection was the tangy and citrusy lemon honey. It completely lived up to all my expectations. Custardlike and creamy, lemon honey could easily be drizzled over fresh berries in a graham cracker crust and garnished with a sprig of mint. Native to southern Italy, this tart honey was as refreshing as the fruit itself. Next, I was introduced to a crystallized honey *romarino* or rosemary. I slathered a spoonful of this luscious creamed honey onto my tongue and could feel the finely granulated texture melting with refreshing sweetness similar to fondant on a wedding cake. Herby and minty, this honey would complement a cup of iced tea. What was there not to love? Francesco's favorite was the eucalyptus honey. He said it was very effective at soothing his children's coughs. Eucalyptus grows wild along the Mediterranean, so there is no shortage of eucalyptus nectar. I could smell the highly aromatic, menthol-like scent before I even dipped my spoon into the jar. Chestnut honey, the darkest of all the honeys, was an Italian favorite and kitchen staple. Its earthy, nutty, somewhat caramelized flavor was definitely an acquired taste. I could imagine this honey being used in a marinade or drizzled over stinky blue cheese.

The vast range of honeys harvested by this beekeeping duo was impressive. Their appreciation of honey was as deep-rooted as their respect for fine wine. I was treated to a walk through their apiary and did not leave without remembering to purchase half a dozen jars of Amalfi honey.

Who would think honey would carry the same culinary prestige as wine and olive oil? But in Italy, honey is so highly respected that it is sold alongside both these culinary favorites in *enotece* or wine shops. The medieval commune of Montalcino is known as *la citta del miele*, or the city of honey. Here, beekeepers harvest some fifty varietals of honey including acacia, chestnut, eucalyptus, wildflower, cherry, thyme, and lemon. They are as proud of their honey as they are of their renowned Brunello wine.

Tasting and Evaluating Wine and Honey

What began as a simple garden hobby to divert my attention from the stresses of everyday work soon turned into a culinary journey. I always remember the quote by Joseph Campbell: "If you follow your bliss, you put yourself on a kind of track that has been there all the while, waiting for you, and the life that you ought to be living is the one you are living."

My bliss was pure honey, the sweetness of a flower blossom that had been kissed by a honeybee. The more I traveled, the more I collected and researched samples of rare and exquisite honey from around the globe, each with its different nectar source, flavor, aroma, color, provenance, and seasonal and tasting notes. Friends brought me honey from their trips abroad, and I even found myself asking strangers from faraway places, whom I'd just met, to send me unusual honeys from their native lands. Each specimen opened a new world of the cultural, geographic, historic, and culinary delights of a particular region.

Not long after my trip to the Amalfi coast, I landed a small design project with an Italian wine importing company. I had the luxury of being trained by a well-known sommelier and importing partner. Joe could blind taste any bottle of wine and tell you its life story, including the grape variety, winery, region of origin, and even what year the grapes were harvested. The sophistication of his palate was staggering. Apparently, there are a chosen few, many of whom are women, who experience taste with tremendous intensity and detail. (Thirty-five percent of all woman, but only 15 percent of men, are known to have superior senses of taste because of an increased number of *fungiform papillae*, which hold our taste buds.) They are called *super tasters*, and they are the perfect candidates to judge and evaluate wine. Joe told me that accomplished wine tasters could recognize and memorize flavors better than most people. He was sure that I could develop my taste memories and a conscious palate after some experience and exposure to many different wines.

Joe explained that the grapes used to make wine could take on different characteristics from year to year depending upon the French concept called *gout de terroir,* or taste of place. Terroir (pronounced *tair wahr*) is the sum of the unique combination of geographic location, soil and its mineral content, climate, annual rainfall, and temperature during a given season, all of which gives each wine its individual profile and personality. These unique aspects of a region influence the personality of a wine, which is why Champagne is the only designated region for Champagne, and Chianti, Italy, for Chianti wine. These traditions are protected by European Union laws and are now being recognized here in the United States. Joe and I tasted wines all made from the same variety of grape grown in different regions, and then different varieties of grapes all grown in the same region. All of these wines tasted remarkably different from one another.

Joe explained the sensory process of evaluating wines: Before you taste it, you look at it and sniff it, evaluating it for color and then swirling it in your glass to release the bouquet. Finally, you swish it in your mouth to release the full flavor and texture. The last taste sensation was a wine's aftertaste or the final impression you are left with. Tasting, he explained, is essentially an olfactory experience. Try tasting something with a stuffed-up nose. It's impossible. Wine also has its own descriptive vocabulary, and Joe always had the perfect term to precisely describe each taste we experienced. *Dry, smoky, fruity, feminine,* or *sweet* were some of the descriptive words he used.

It quickly became apparent to me that there were many correlations between wine and honey. Both were agricultural products harvested in the late summer, and the tastes of both were governed by floral sources. I had collected a number of different honeys from around the world and started taking note of the differences in color, texture, flavors, and nectar sources. The varietals of honey that I'd sampled from around the world seemed endless, and soon I convinced Joe to put his super palate to use tasting honey. Many of the same terms used to describe wine could also be applied to the taste of honey, and we made up a few terms of our own. At this point I began to understand the real depth of experience that honey had to offer. My experience of working with wine made my reverence for honey even deeper, and I believed that honey deserved the same noble recognition as wine.

Artisanal Honey, Terroir, and Vintages

It can be said that honey is only as good as the beekeepers who harvest it.

Artisanal honeys are those produced by individuals using traditional methods and thus preserving the integrity of their products. With artisanal honey, quality and character are highlighted, rather than quantity and consistency. Beekeepers have to make many decisions regarding the management of their honeybees during a single season. Timing is everything, so colonies have to be at their peak strength and available to forage the fields at the exact time of the nectar flow. Beekeepers must select appropriate field locations for their honeybees and know when the nectar flow begins, when to add and remove honey shallows, and the best procedure to extract the honey.

Mother Nature must cooperate too. The flavor profile and essence of any bottle of honey will depend largely on its terroir, which fundamentally determines the type of plant life in a specific growing area. Plants will bloom and produce nectar according to climate, elevation, weather and rainfall, nutrients in the soil, temperature, and available sunlight. Each flower has its own distinct type of nectar, which is produced at specific times of the season, in specific regions, and more complexly, at specific times of the day. Ultimately, the type of nectar and the flowers dictate the composition of a honey, like the grape dictates the resulting wine. Honey harvests can have "good" years and "bad" depending upon the climatic diversity. Changes in weather dictate which years or vintages were bountiful. Limited harvest honeys or honey from a rare nectar source can command higher prices than commonly available honeys. In the near future, beekeepers will begin identifying these vintages and provenances on their honey labels.

A honey's flavor profile will vary not only from year to year but also from hive to hive. The exact same field of flowers, if produced by two different hives that stand right next to each other, can bear honey that is completely different in color, flavor, texture, and aroma. Blueberry-blossom honey from Maine might be lighter in color than a blueberry-blossom honey from Michigan. A clover honey harvested from a hive in 2006 could have a slightly different flavor profile from the clover harvested from that same hive in 2007. There are certain locations around the world that hold prestige when it comes to making honey. Georgia is known for its tupelo honey, Hawaii for its kiawe, France for its lavender, and Australia for its manuka. Honeybees foraging in these specific provenances harvest nectar only from these plants. I refer to these honeys as *single origin*, meaning from a specific region.

In Europe, there are certifications granted to protect agricultural products that are produced in respected geographic regions. These certifications are a guarantee that the products produced in these specific regions follow stringent rules and traditions with regard to manufacturing, purity, and origin. In France there is the *Appellation d'origine controlee* (AOC) In Italy the *Denominazione di Origine Controllata e Garantita* (DOCG) and in Spain the *Denominación de Origen* (DdO). Food and wine are commonly granted these certifications and recently, honey has begun to be certified as well.

Honey from the island of Corsica was the first to be granted this respected AOC status. Presently, there are six certified varietals of Corsican honey: *Printemps, Maquis de printemps, Miellats du maquis, Châtaigneraie, Maquis d'été,* and *Maquis d'automne. Miel de la Alcarria*, produced near Guadalajara, Spain, has also been granted the DdO certificate.

Also in France, the *label rouge* (red label) guarantees that a honey product is of the highest quality available. Lavender honey from Provence, as well as that region's fennel, thyme, and chestnut honeys, have been granted the *label rouge*.

TASTING AND JUDGING HONEY AND HONEY PRODUCTS

How do we judge the quality and flavors of a honey? I found answers to this question in the center of London, where I attended the National Honey Show, a three-day event complete with lectures, foods, crafts, books, equipment, and paraphernalia all relating to honey and honeybees. One large room was dedicated to the judging of honey samples from all over the world. The entries were meticulously lined up on endless rows of white-tiered tables. Beekeepers had painstakingly prepared and imported their honey entries from their home countries according to the show's strict guidelines.

Decorated honey judges, who spoke of honey by its color, texture, and aroma, had their own official tasting vocabulary and even scorecards ready for use. These judges were trained and certified by the strict Welsh method. Not only did they judge honey but they also tasted meads and critiqued beeswax candles, arts, and crafts and anything else having to do with the honeybees. Each judge was accompanied by a steward, a judge in training, who takes notes on the judge's observations. Judges use a special tool called a *refractometer* to determine the water content of a honey, and a set of colored glass plates to identify exact color grades. Sometimes a flashlight is used to observe the clarity of the jar or to see if there are any air bubbles or pollen particles afloat in the honey. A single

Single-Varietal Honeys

Honeybees tend to visit a single flower species on any one foraging trip; examining pollen as the bees return to the hive has proved this constancy. As long as the source flower is in bloom, bees will continue to gather nectar from that flower and make honey from this single flower. This loyalty lasts for varied lengths of time and creates single-varietal honeys (also called, monofloral or unifloral honeys).

The botanical and geographic provenance of any bottle of honey can be determined through the study of its pollen sediments. All geographical areas have their own specific flora and fauna associations, and terroir plays a part in these associations. To confirm the provenance of a particular bottle of honey, you have to identify the combinations and characteristics of pollen sources particular to that region. A honey's entire biography is right there in the bottle.

trace of honey on the inside of the lid will disqualify a perfectly prepared jar of honey—no exceptions. All of the entries were sorted out by specific classes, and placing each entry in its correct category was of the utmost importance, or a perfect entry could be disqualified. First-place winners were awarded a humble dime-sized sticker with a red star.

The second-largest category of entries was the beeswax products. Cleaning and properly preparing your beeswax was apparently an art unto itself, and the English beekeepers took this art seriously.

A favorite category for the judges was undoubtedly the tasting of the meads. There were as many types of mead as there were honey.

Meads made with fruits, spices, and herbs were all on display and back-lit to emphasize their wide variations in color. The area of *handmade* entries exhibited unconventional crafts illustrating anything to do with honeybees. On exhibit were collectible honeypots made of delicate porcelain, clay, and ceramic and molded into the shapes of skeps, honey bears, and honeybees. I encountered many original works of art that were painted with beeswax. This painting technique, called *encaustics*, was used more than three thousand years ago in ancient Rome and Egypt, but was made famous by classical Romano-Egyptian Fayum portraits, which today hang in the Metropolitan Museum of Art. Being an artist, I was absolutely smitten with the process of melting of beeswax, mixing it with colored pigments, and painting with it. At future beekeeping events I would watch demonstrations and even try my hand at a painting. For me, encaustic painting is where beekeeping meets art.

One judge told me that the English beekeepers are envious of how many different floral sources we have in the United States, but to my great surprise, I was only one of five Americans in attendance that session.

Once back home, I sought out a honey-judging certification at the University of Georgia so that I could learn more about the technical aspects of appreciating honey. Soon, it became clear to me that people all over the world knew about honey and bees, and it became my mission to spread the good word about honey in this country.

GRADING OF HONEY

In the United States commercial honey is graded by the U.S. Department of Agriculture (USDA) according to a standard system. These standards are based upon a technical point system that accounts for water content, flavor and aroma, clarity, and absence of defects. Similar to the grading used for maple syrups, these grades are listed mostly on commercial honeys found in grocery stores rather than artisanal honey purchased at a farm stand or your local beekeeper. The standards are Grades A, B, C, and Substandard. The most desirable is A, and the least, of course, is Substandard.

Grade-A honeys are clear, have good flavor and aromas, and are free from crystallization, air bubbles, pollen, propolis, and wax particles. Grade-B honeys are considered to be reasonably clear, to have reasonably good flavor and aromas, and may have few air bubbles, pollen, propolis, and wax particles. Grade-C honeys are fairly clear, have fairly good flavor and aromas, and are fairly clear of air bubbles, pollen, propolis, and wax particles. Substandard is extracted honey that fails to meet the requirements of USDA Grade C.

THE VISUAL PROPERTIES OF HONEY

The first thing you will notice when looking at honey through a clear glass jar is its clarity or visual properties. Some of the honeys I have collected appear cloudy or opaque, while others are crystal clear. Cloudiness can be the result of air bubbles, pollen grains, or other fine particles or materials floating in the bottle and does not necessarily mean the honey is not tasty or acceptable. During the process of extracting honey from the frames, air can get into the honey, producing a foamy layer on the top. Foam or froth in honey is unavoidable. It usually rises to the top of the honey jar and will eventually disappear after the honey settles. For cosmetic purposes, excess foam should be removed with a spoon or a strainer. When describing a honey's relative visual properties a few terms are used: *clear*, *reasonably clear*, and *fairly clear*.

THE COLOR OF HONEY

The color of honey is determined by its floral source and mineral content. Honey color varies naturally and comes in a wide range of tonalities, from water white or clear to light yellow, gold to amber, purple to dark amber, and in extreme cases, black. I have seen red honeys and even ones with a greenish tint. The best way to judge a honey's color is by filling a small white cup about ⅛ of an inch (3 mm) with a sample. This way you can see the color of the honey sample against the white background of the cup. Use a designated color fan called a *Pfund color grader* to compare the color of your honey with the colors on the fan. Simply match the honey sample with the closest value. There is also another system that uses colored glass tiles with designated honey colors. The seven designated colors of honey are water white, extra white, extra light amber, light amber, amber, dark amber, and dark. Although there are variations on these accepted colors, all honeys can be categorized in one of these seven colors. All honey tends to deepen in color as it ages, but this change does not affect its flavor.

Aroma or Nose of Honey

The fragrance or bouquet of a honey refers to its aroma or nose. Of our five senses, our sense of smell is approximately a thousand times more sensitive than our sense of taste, and humans can detect 10,000 different odors but only five flavors. About 80 percent of our taste buds are located in our noses, so most of our knowledge of what a honey tastes like actually comes from our noses before we even experience the honey on our tongues. What we may call flavor is roughly 75 percent smell (olfaction) and 25 percent taste (gustation). Each honey has a variety of aromas depending upon its temperature and its predominant floral source. Some common words used to describe these aromas are *flowery, fruity, spicy, putrid, resinous,* and *burned*. The overall odor intensity can be rated on a scale of one to nine; one is low intensity, and nine is high intensity. Water is the best palate cleanser between tasting different honeys.

Many factors will affect the way you perceive a honey—or any food for that matter—and therefore the degree to which you will appreciate that honey. These factors include temperature of the room, time of day, degree of lighting, utensils, expectations, personal health, and even air currents. Many flavors cannot be correctly accessed when the honeys are too cold or hot.

Texture, Viscosity, and Body of Honey

The feel of a honey in your mouth can be described as its *texture* or *viscosity*. Words for describing honey texture are *smooth*, *slippery*, *gritty*, *velvety*, *creamy*, *buttery*, *thick*, *thin*, *watery*, *drippy*, *grainy* and *granular*, and *runny*. The body of honey can range from *watery thin* to *light*, from *thick* or *heavy* to even *oily*. These words express the sense of the honey's feel and weight on your tongue.

Temperature and moisture content are important factors that can change all of these qualities dramatically. Heat tends to make honey pour more quickly, giving it a thinner texture, and cooler temperatures can make honey stiff. Keeping honey at room temperature prior to tasting is the best way to appreciate and savor its flavor.

Too much moisture in a jar of honey can cause early fermentation. There are a few ways to judge the amount of moisture in a jar of honey. The first is to observe how quickly a bubble of air rises to the bottom of the jar as you turn it upside down. If the air bubble floats up very quickly, this could mean a watered-down honey. The second method is a scientific approach using a tool called a refractometer. This handheld tool looks like a miniature telescope. Place a dab of honey on its lens, and you can accurately measure moisture content. An acceptable level of moisture in honey is 18 percent.

How to Taste Honey

Honey is said to have "good flavor" when the floral source is clearly recognizable, and the honey is free from fermentation, bitterness, or burned flavors. Tasting is very different from merely swallowing, and tasting honey is an exercise in comparing and contrasting flavors, textures, aromas, and color. Our taste buds can distinguish between the following five flavor sensations: sweet, sour/tart, salty, bitter, and *umami*, which is that sense of savoriness. The taste buds for each flavor are located at different areas on the tongue; for this reason, it is recommended that you move the honey around your mouth as you taste it. Tasting more than one honey at a time provides a context in which to compare tasting notes. A range of honeys should be tasted in the same fashion as wines, beginning with the lightest in color, moving on to medium ambers, and then finishing with the darkest. Light-colored honeys typically have a mild flavor, while dark-colored honeys are usually stronger in flavor. Tasting has many levels: the first impression, the actual taste, and the finish. Take a sip of water in between each sample to clear your palate. A nice piece of crusty bread and a variety of cheeses can complement the tasting experience.

A honey tasting ideally should be held during the honey harvest season in late summer or early autumn, since this is when most beekeepers have taken the fresh honey shallow supers off their hives. If you travel and purchase honey from local beekeepers, they are usually happy to give you a sampling of their treasure.

When tasting honeys, the terms on the next page will come in handy as you discuss and describe the experience. Begin by drizzling a sample of honey onto your tongue, and let it melt for a few seconds. Spread it around your mouth while thinking about its body and bouquet. Is it woody, floral, full, light, crisp, buttery, well balanced, overly acidic? Does it have a long finish or an abrupt end? Do you like it or hate it? Each person has his or her own individual taste and predisposition to flavors. Remember, there is no right or wrong, and you should never eat something you dislike. Although most people enjoy something sweet, I have found that certain individuals genuinely dislike honey.

THE RED BEE LANGUAGE OF HONEY

Here is a vocabulary list of analogies I use to describe the many tastes of honeys:

VEGETATIVE: hay, straw, wheat, green tea, fresh-cut grass, green bananas

HERBAL: camphor, menthol, peppermint, rosemary, eucalyptus

CARAMELIZED: molasses, caramel, burnt, toffee, brown sugar, maple

BUTTERY: melted butter, smooth, butterscotch, silky, froth, rich

BITTER: tart, tangy, crisp, natural flavors of some nectars

FRUITY: bright, citrusy, berries, tropical, dried fruits

> *Citrus:* orange, lemon, tangerine
>
> *Berry:* blackberry, raspberry, strawberry, black currant
>
> *Tree fruit:* cherry, plum, apple, peach
>
> *Tropical fruit:* mango, pineapple, apricot
>
> *Dried fruit:* prune, fig, raisin

SWEET: sugary, saccharine, syrupy, candy, vanilla

Honey generally tastes sweeter than white sugar to most people's palates, but with some experience you will learn to enjoy the layers of flavors in honey.

SALTY: briny, saline, brackish, salted

NUTTY: walnut, hazelnut, almond, pecan, pine

WOODY: rustic, vanilla, cedar, oak, smoky, leathery, coffee, chocolate, tobacco

EARTHY: dusty, rich soil, musky, black tea

PUNGENT: yeasty, vinegar, sour, acid, biting, fermented (Honey in its fermented state can taste yeasty.)

FLORAL: perfumy, rose, violet, geranium

SPICY: prickly, sharp, cloves, nutmeg, cinnamon, pepper, licorice, anise

WAXY: soft, chewy, flavor of beeswax

METALLIC: iron

Honey Pairing

Artisanal honey can make just about any food taste divine because it naturally pairs well with all food groups, but especially dairy products. Fruits, nuts, breads, and especially wine and cheese make unforgettable accompaniments. An easy way to determine which types of honeys you will appreciate is by learning more about the types of foods and wines you already enjoy. There are two ways to think about pairing honey with food: The first is to find a honey that is similar in taste and flavor to the foods you are preparing. The other is to contrast the flavors of the honey with the flavors of the foods. When you are pleased by the sensations in a pairing, the honey and food is considered a harmonious combination. In Italy honey is paired with the country's famed pecorino and ricotta cheeses; in Spain it's paired with tapas, in Greece with yogurt, and in Israel with apples and challah bread. So if cooking and travel are your pleasures, it won't be long before artisanal honey finds a place in your kitchen.

Honey in the Kitchen

Pure honey adds multidimensional tasting notes when added to foods, whereas sugar simply adds sweetness. On the average, honey is one to one and a half times sweeter than sugar, so you will tend to eat less honey. One tablespoon of table sugar contains forty-six calories, while one tablespoon of honey has sixty-four calories.

Honey acts as an emulsifier in salad dressings, helping to keep oil and vinegar mixed. Honey will slide smoothly off measuring utensils if you lightly coat the utensils with vegetable oil or cooking spray beforehand.

TIPS FOR BAKING WITH HONEY

Honey is also a humectant, a substance that either absorbs or helps other substances absorb moisture. Another scientific term for this property is *hygroscopic*, meaning it can actually attract moisture. For bakery foods, more moisture generally means longer freshness or shelf life. In other words, treats baked with honey tend to be moister and to stay moist longer.

When using honey in a recipe, reduce the amount of any other liquid by ¼ cup for each cup of honey used and add ½ teaspoon of baking soda for each cup of honey used. Reduce the temperature of the oven by 25°F to prevent overbrowning.

STORING YOUR HONEY

Many folks are amazed when I tell them that pure honey *never* needs refrigeration and *never* spoils if stored correctly. The optimum storage temperature for honey is between 70 and 80°F (21 and 27°C), and it should be stored in airtight containers.

Appendixes

Deciphering a
Honey Label

Whether you are traveling or visiting a gourmet food shop here in the United States or abroad, you will always find interesting honeys for sale. You don't need to spend a lot of money to experience a rare or unusual variety of honey.

Some people will choose their honey by their labels. Beginners may opt for a label with a sweet honeybee, and experienced honey lovers may take a chance on an esoteric label. Most honey labels will be written in English, but for the fancy imported ones you may come across, here is an alphabetic listing of honey words and terminology with translations in the most popular honey-producing languages: English, Italian, French, and Spanish. This list should help get you out there experimenting with artisanal honeys. If you happen to find a special jar of honey whose label you cannot read, but it still looks interesting, my suggestion is to buy the bottle. You never know what type of honey might become your favorite.

ENGLISH	ITALIAN	FRENCH	SPANISH
almond	mandorlo	amandier	almendro
apiarist (U.S.)	apicoltore	apiculteur	apicultor
apiary	apiario	rucher	colmenar
apple	melo	pommier	manzano
apricot	albicocco	abricotier	albaricoquero
baking honey	miele di pasticceria	miel de pâtisserie	miel de pastelería
bee	ape	abeille	abeja
beehive	arnia	ruche	colmena
beekeeper	apicoltore	apiculteur	apicultor
beekeeping	apicoltura	apiculture	apicultura
black currant	ribes nero	cassis	grosellero negro
blackberry bramble	rovo	ronce	zarzamora
blackthorn	prugnolo	épine-noire	endrino
broad bean	fava	fève	haba común
buckwheat	grano saraceno	sarrasin	trigo sarraceno

ENGLISH	ITALIAN	FRENCH	SPANISH
centrifuged honey	miele estratto	miel extrait	miel extraída
cherry (cultivated)	ciliegio	cerisier	cerezo
chestnut, horse	ippocastano	marronnier d'Inde	castaño de Indias
chestnut, sweet	castagno	châtaignier	castaño
chunk honey	miele in favo	rayon de miel	trozos de panal con miel
clear honey	miele liquido	miel liquide	miel liquida
clover	trifoglio	trèfle	trébol
clover, crimson	trifoglio incarnato	trèfle incarnat	trébol incarnado
clover, red	trifoglio rosso	trèfle rouge	trébol rojo
clover, sweet	meliloto	mélilot	meliloto
clover, white	trifoglio bianco	trèfle blanc	trébol blanco
comb honey	miele in favo	miel en rayon	panal de miel
creamed honey	miele cremoso	miel crémeux	miel crema
currant, black	ribes nero	cassis	grosellero negro

ENGLISH	ITALIAN	FRENCH	SPANISH
currant, red or white	ribes rosso	groseillier	grosellero de racimos
dandelion	tarassaco	pissenlit	diente de líon
dark honey	miele scuro	miel foncé	miel oscura
dark amber honey	miele ambra	miel ambré	miel ámbar oscuro
eucalyptus gum	eucalipto	eucalyptus	eucalipto
extra light amber honey	miele ambra extra chiaro	miel ambré extra-clair	miel ámbar extra claro
extracted honey	miele estratto	miel extrait	miel extraíd
flavor	gusto, sapore	goût, saveur	gusto
fructose	fruttosio	fructose	fructosa
fruit sugar	zucchero di frutta	sucre de fruit	azúcar de frutas
gallberry	ilex glabra	houx	ilex glabra
golden rod	solidago	verge d'or, solidage	vara de oro
gooseberry	uva spina	groseillier	grosellero espinoso

ENGLISH	ITALIAN	FRENCH	SPANISH
to grade (honey)	classificare	classer le miel	clasificar
grading glass	colorimetro	colorimètre	colorímetro
granulation	granulazione	granulation	granulación
hawthorn	biancospino	aubépine	espino albar
hazel	nocciolo	coudrier, noisetier	avellano
heather	erica comune	callune	brezo común
heather, bell	erica cinerea	bruyère cendrée	brezo ceniciento
heather honey	miele di erica	miel de callune, miel de bruyère	miel de brezo
hive, beehive	arnia	ruche	colmena
hive bee	ape mellifera	abeille domestique	abeja doméstica
holly	agrifoglio	houx	acebo
hollyhock	malva rosa	rose trémière	malva real
honey	miele	miel	miel
honeybee	ape mellifera	abeille domestique	abeja melífera

|---|---|---|---|
honeycomb	miele in ifavo a pezz	rayon de miel	panal de miel
honeydew	melata	miellat	mielato
honeydew honey	miele di melata	miel de miellat	miel de mielato
honey grader	rifrattometro	colorimètre à miel	colorímetro
honey, granulated	miele cristallizzato	miel granulé	miel granulada
honey harvest	raccolta del miele	récolte du miel	cosecha de miel
honey house	mieleria	miellerie	mielería
honey label	etichetta per il miele	étiquette à miel	etiqueta para miel
honey plant	flora nettarifera	plante mellifère	planta melífera
honey show	mostra di miele	exposition de miel	exposición de miel
lavender	lavanda	lavande	espliego
light amber honey	miele ambra chiaro	miel ambré clair	miel ámbar claro
light-colored honey	miele chiaro	miel clair	miel clara

ENGLISH	ITALIAN	FRENCH	SPANISH
lime or linden	tiglio	tilleul	tilo
liquid honey	miele liquido	miel liquide	miel líquida
locust tree	robinia pseudoacacia	faux acacia	falsa acacia
loosestrife, purple	riparella	salicaire	salicaria
mead	idromele	hydromel	hidromiel
mustard	senape	moutarde	mostaza
nectar	nettare	nectar	nectar
nutrition	nutrizione	nutrition	nutrición
orange	arancio	oranger	naranja
over-heated honey	miele surriscaldato	miel surchauffé	miel sobrecalentada
peach	pesco	pêcher	melocotonero
pear	pero	poirier	peral
pine, Scots	pino silvestre	pin sylvestre	pino Silvestre
plantain	piantaggine	plantain	plátano
plum	prugno	prunier	ciruelo

ENGLISH	ITALIAN	FRENCH	SPANISH
pollen grain	granulo pollinico	grain de pollen	grano de pollen
pressed honey	miele torchiato	miel de presse	miel extraída por presión
propolis	propoli	propolis	propóleos
queen	regina	reine	reina
rape, coleseed, or colza	ravizzone	colza	colza
raspberry	lampone	framboisier	frambueso
red currant	ribes rosso	groseillier à grappes	grosellero de racimos
refractometer	rifrattometro	réfractomètre	refractómetro
rosemary	rosmarino	romarin	romero
royal jelly	pappa reale	gelée royale	jalea real
sage	salvia	sauge	salvia
to store (honey)	immagazzinare	amasser	amasar
strainer (for honey)	filtro per il miele	filtre à miel	filtro para miel
sunflower	girasole	tournesol	girasol

ENGLISH	ITALIAN	FRENCH	SPANISH
syrup	sciroppo	syrop	jarabe
taste, sense of	gusto	goût	gusto
thyme	timo	thym	tomillo
tulip tree	tulipifero	tulipier	tulipero
tupelo	tupelo	tupélo	tupelo nisa
variety	varietà	variété	variedad
vetch	veccia	vesce	veza
water-white honey	miele bianco acqua	miel blanc d'eau	miel blanco agua
wax	cera	cire	cera
white honey	miele bianco	miel blanc	miel blanca
willow	salice	saule	sauce

75 Varietals of Honey

1. ACACIA

PLANT CHARACTERISTICS: From the black locust tree, which has fragrant, pea-shaped white or yellow clustered flowers producing gray-colored pollen.

BLOOMS: May or June.

BOTANICAL NAME: *Robinia pseudoacacia*

COMMON NAMES: False or white acacia; honey locust; black, white, or yellow locust; *Robinier, Robinia*.

PROVENANCE: Native to the U.S. Appalachian Mountains and the Ozark Mountains. It can also be found in Hungary, the United Kingdom, Canada, France, Italy, and China.

TERROIR: Prefers humid climates and sandy, well-drained soil types. Found on prairies and fields in sunny locations.

HONEY COLOR: Water white, absolutely clear with a shiny, glasslike appearance.

TASTING NOTES: A light, delicate, and flowery honey. Primary notes of dried pineapple, vanilla, almond, and butterscotch, with subdued caraway notes. Slightly acidic. Heavy body. A fairly high fructose content allows this honey not to crystallize quickly.

PAIRINGS: Drizzle over Pecorino Romano, serve with fresh figs, apricots, marcona almonds, and chardonnay or prosecco.

2. ALFALFA

PLANT CHARACTERISTICS: A flowering herb and an important honey plant with white, greenish yellow or violet flowers and lemon yellow pollen. Yields abundant amounts of honey when fields are left uncut and grown for seed.

BLOOMS: April to October.

BOTANICAL NAME: *Medicago sativa*

COMMON NAMES: Lucerne, Luzerne, Lucerne grass, Spanish trefoil, purple medick, *Erba medica*.

PROVENANCE: Native to Europe and China. The alfalfa belt of the United States is found in the northwest part of the country.

TERRIOR: Prefers well-drained, fertile lime soils. Grows in a wide range of climates, from cold northern plains to mountains to deserts.

HONEY COLOR: Light amber with a warm orange tint.

TASTING NOTES: A smooth, rich, buttery aftertaste. Mild tones of minty herbs, fresh grass, and straw. Slightly metallic. Beeswax aroma. Full body. Granulates quickly. Alfalfa honey is high in protein.

PAIRINGS: Drizzle over robiola or classic blue cheeses on savory herbed bread, and serve with chardonnay or pinot blanc.

3. APPLE BLOSSOM

PLANT CHARACTERISTICS: A fruit tree with fragrant, pinkish white clusters of flowers producing pale yellow pollen. Must be cross-pollinated by insects, and honeybees are the best pollinators for the job.

BLOOMS: Early spring.

BOTANICAL NAME: *Malus domestica*

COMMON NAMES: *Melo, pommier, manzano.*

PROVENANCE: Native to Asia. Found in the United States, especially Washington, New York, Michigan, California, Pennsylvania, and Oregon.

TERRIOR: Prefers temperate climates with fertile, sandy soils. It is best adapted to areas in which the average temperature approaches or reaches freezing during at least two months. Cool weather lengthens bloom periods. Requires plenty of sunlight.

HONEY COLOR: Light golden brown.

TASTING NOTES: Delicate, crispy apple flavor. Tart and astringent. Granulates quickly.

PAIRINGS: Drizzle over cheddar cheeses and sliced fresh apples on ginger snaps, and serve with Riesling, champagne, or mead. Add to glazes for pork chops or sausages.

4. ASTER

PLANT CHARACTERISTICS: A perennial plant with erect or spreading leafy stems with daisylike, deep golden yellow or blue-violet flower heads in branched clusters. This is a rare honey because it is often mixed with goldenrod.

BLOOMS: September through November.

BOTANICAL NAME: *Aster spp*

COMMON NAMES: Chrysanthemum, starflower.

PROVENANCE: Native to the midwestern and eastern United States, as well as Canada and Europe.

TERRIOR: Grows in wet marshes near woody regions with full sun. Some varieties thrive in dry, sandy soil.

HONEY COLOR: Water white to light amber.

TASTING NOTES: Can have an off aroma until it ripens. Spicy overtones. Firm, chewy texture. Firm enough to eat straight from the jar, like candy. Granulates quickly into fine grains, making it difficult to extract from the hive.

PAIRINGS: Spread over cornmeal and spice breads

5. AVOCADO

PLANT CHARACTERISTICS: Tree or shrub with greenish yellow flowers with deep yellow to brown heavy and sticky pollen. Avocado trees bear an egg-shaped fruit, which is technically a berry. An important honey and nectar plant for honeybees.

BLOOMS: Spring and again in summer. Grown mostly for fruit.

BOTANICAL NAME: *Persea americana*

COMMON NAMES: Alligator or butter pear, midshipman's butter, vegetable butter, *avocatier, aguacate, aguagate, palta.*

PROVENANCE: Native to California, Florida, Mexico, the Caribbean, Israel, Chile, Argentina, and many African countries.

TERRIOR: Survives in diverse soils, red clay, sand, volcanic loam, lateritic soils, limestone. Requires humidity and a tropical or near tropical climate, especially during flowering and fruit setting.

HONEY COLOR: Very dark amber.

TASTING NOTES: Robust, spicy aroma. Rich flavor with hints of molasses and caramelized sugar. Heavy body. Slow to granulate. This honey is full of minerals and vitamins.

PAIRINGS: Drizzle over Emmental or Gruyére cheeses and sliced almonds, and serve with dessert or port wines. Pour over pancakes, waffles, or chocolate ice cream.

6. BASSWOOD

PLANT CHARACTERISTICS: Flowering tree with clusters of yellow or cream-colored, very fragrant flowers yielding large quantities of nectar very quickly. A popular honey source in the eastern United States where it is called the bee tree.

BLOOMS: Late June and July.

BOTANICAL NAME: *Tilia americana*

COMMON NAMES: American linden, lime, linden tree, white or American basswood, *Tilleul, Tilia.*

PROVENANCE: Native to northeast and central United States and Central America.

TERRIOR: Prefers deep, moist soils in limestone regions, and cold winters and warm summers. Humid air promotes nectar secretion.

HONEY COLOR: Water white to extra white amber.

TASTING NOTES: Warm aromatic, herbal notes. Spicy, green ripening fruit. Sweet, biting, astringent flavor with a strong medicinal finish.

PAIRINGS: Drizzle over goat cheeses and sliced green apples, and serve with chardonnay and sauvignon blanc. Mix into honey mustards and squash soups.

7. BLACKBERRY

PLANT CHARACTERISTICS: A prickly shrub with pale pink or white-lavender flowers with clear, dull, greenish white pollen. Requires pollination by the honeybee.

BLOOMS: Late spring and early summer.

BOTANICAL NAME: *Rubus fruticosus*

COMMON NAMES: Bramble, dewberry, goutberry, *rovo, ronce, zarzamora.*

PROVENANCE: Native to Europe, Asia, and Africa, as well as Oregon, Washington, and Mexico.

TERROIR: Tolerates poor soil wastelands, woods, and hillsides.

HONEY COLOR: Ashy or smoky light to extra light amber.

TASTING NOTES: Delicate, sweet, fruity, deep raspberry flavor. Full body. Slow to granulate.

PAIRINGS: Drizzle over sharp white cheddar and sliced fresh peaches, and serve with Syrah. Great for making jam jellies and preserves. Serve over pancakes and corn muffins. Used in wine making.

8. BLACK MANGROVE

PLANT CHARACTERISTICS: An evergreen shrub with small white flowers with yellow centers.

BLOOMS: May to July.

BOTANICAL NAME: *Avicennia germinans*

COMMON NAMES: Blacktree, blackwood, mangrove, *mangle negro, courida.*

PROVENANCE: Native to Belize and Ambergris Caye. Found on the the banks of the Indian River Lagoon and Gulf Coast of Florida, Louisiana, Mississippi, and Texas.

TERROIR: Thrives in the shallow, muddy, brackish wetlands of tropical regions. Requires full sun.

HONEY COLOR: Very light amber with a greenish tint.

TASTING NOTES: Thin bodied. Swampy aroma, with a mild, sweet, but brackish flavor. Granulates rapidly.

PAIRINGS: Drizzle over Camembert and sliced fresh pineapple, and serve with chardonnay. Mix with brines for pickles and sauerkraut.

9. BLUEBERRY

PLANT CHARACTERISTICS: A flowering shrub with bell-shaped, white or pinkish flowers. The fruits are edible when they change from pale green to deep blue at their peak in July. Blueberries are self-pollinating, but if honeybees cross-pollinate with another plant, the fruit will be larger, ripen earlier, and have fewer seeds. Blueberries are a source of antioxidants, as is the honey made from blueberry flowers.

BLOOMS: March.

BOTANICAL NAME: *Vaccinium corymbosum*

COMMON NAMES: Highbush berry.

PROVENANCE: Eastern United States, especially Maine, Massachusetts, and Rhode Island, but also Michigan, Oregon, Washington, and Canada.

TERRIOR: Requires moist, acidic soil and full to partial sun.

HONEY COLOR: Cloudy, light amber.

TASTING NOTES: Buttery and smooth texture. Fruity with hints of green leaves and fresh lemon. Granulates quickly into large crystals.

PAIRINGS: Mix into vanilla yogurt with chopped walnuts and sliced fresh bananas. Drizzle over sour cream coffee cake or crumb cake.

BLUE CURLS

10. BLUE CURLS

PLANT CHARACTERISTICS: Strong evergreen herb with aromatic green leaves. Blue, densely clustered flowers with loden green pollen.

BLOOMS: August to first frost.

BOTANICAL NAME: *Trichostema lanatum*

COMMON NAMES: Woolly blue curls, Romero, California rosemary, American wild rosemary.

PROVENANCE: Native to California's Fresno and Ventura counties.

TERRIOR: Well-drained, dry slopes and plains of the coastal regions. Requires full sun.

HONEY COLOR: Extra light yellow-amber to milky white.

TASTING NOTES: Minty, fruity flavor. Granulates quickly and smoothly.

PAIRINGS: Drizzle over goat cheese and cashews, and serve with merlot or Riesling. Add to glaze for lamb or lemon chicken. Used in honey butter.

11. BONESET

PLANT CHARACTERISTICS: A medicinal herb with white or purple clusters of flowers; aromatic leaves.

BLOOMS: Late July to October, at the same time as goldenrod, sunflowers, and aster, making pure boneset honey difficult to harvest.

BOTANICAL NAME: *Eupatorium perfoliatum*

COMMON NAMES: Thoroughwort, Indian sage, agueweed, joe-pye, wild sage. (*Boneset* refers to the superstitious practice of wrapping the leaves inside bandages to set broken bones.)

PROVENANCE: Alabama, Tennessee, Kentucky, Louisiana, and Florida.

TERRIOR: Swamps, low damp areas, and pastures.

HONEY COLOR: Very dark reddish amber.

TASTING NOTES: Rank in aroma, but a mellow, herbal, sagelike flavor. Thick, heavy, molasseslike texture.

PAIRINGS: Drizzle over Roquefort cheese on sweet crackers, and serve with champagne, prosecco, or Riesling. Mix with chopped pistachios and roughly chopped fresh cranberries as a glaze for venison.

12. BORAGE

PLANT CHARACTERISTICS: A hardy herb with brilliant blue flowers in loose clusters and with gray pollen. Leaves are silvery, fuzzy, and coarse.

BLOOMS: April to October in North America and November to March in New Zealand.

BOTANICAL NAME: *Borago officinalis*

COMMON NAMES: Viper bugloss, blue borage, starflower, *Soagem, borragine, bourrache, borraja.*

PROVENANCE: A plant new to America but common in the United Kingdom and the Clarence Valley in Central Marlborough, New Zealand.

TERRIOR: Dry wasteland; grows wild in arid areas. Prefers rich soil and full sun.

HONEY COLOR: Medium to dark amber, with a gray tinge.

TASTING NOTES: Herbal and floral bouquet with hints of cucumber and orange pekoe tea. Sugary aftertaste. Delicate and silky texture. Slow to crystallize.

PAIRINGS: Spread over scones and biscuits. Mix into dressings for green and fruit salads. Stir into herbal teas.

13. BUCKWHEAT

PLANT CHARACTERISTICS: An annual herb that is cultivated for flour and is also an important honey plant. The small, fragrant, clustered red flowers secrete nectar early in the morning, and the bees work it intensely, then become very cross in the afternoon when nectar flow ceases.

BLOOMS: Spring until summer.

BOTANICAL NAME: *Fagopyrum esculentum*

COMMON NAMES: *Grano saraceno, sarrasin, trigo sarraceno.*

PROVENANCE: Mainly grown in New York, Pennsylvania, California, Minnesota, Virginia, Michigan, Washington, Ohio, Wisconsin, and eastern Canada. Native to Asia.

TERRIOR: Grows best in cool, moist climates, preferring light and well-drained soils, although it can thrive in highly acidic, low fertility soils as well.

HONEY COLOR: Dark purple turning to deep wine red to black; opaque.

TASTING NOTES: Pungent and somewhat earthy with notes of burnt molasses, cherry, tobacco, and plum. Malty. Buckwheat honey has been found to contain antioxidant compounds, and it is also rich in iron.

PAIRINGS: Drizzle over blue and other strong cheeses, serve with cabernet sauvignon, port, or Madeira. A perfect replacement for maple syrup on pancakes, waffles, buttered corn bread, and gingerbreads. Good for mixing in barbecue sauces and for brewing dark ales.

14. CAROB SEED

PLANT CHARACTERISTICS: A leguminous evergreen shrub or tree with edible seed pods that are a substitute for chocolate. Small, numerous flowers are arranged in a spiral.

BLOOMS: September and October.

BOTANICAL NAME: *Ceratonia siliqua*

COMMON NAMES: St. John's bread. (According to some Christian traditions, St. John the Baptist subsisted on carob seeds in the wilderness.)

PROVENANCE: Native to the Mediterranean region. Found especially in the Atlas mountain range in Morocco, as well as in southern Portugal, Spain, Malta, Sicily, Greece, and Sardinia.

TERROIR: Well-drained, loamy soil.

HONEY COLOR: Dark autumn honey.

TASTING NOTES: Bittersweet honey flavor woven with strains of caramel and subtle chocolate. This honey is an exotic treasure.

PAIRINGS: Drizzle over blue cheeses with hazelnuts and fresh raspberries, and serve with port or muscat dessert wines. Spread on graham crackers or over vanilla ice cream.

15. CAT'S CLAW

PLANT CHARACTERISTICS: A thorny shrub with clusters of bristle-shaped pale yellow flowers. Its prickly spines give it the name *cat's claw*. Important honey plant in the southwest United States.

BLOOMS: Twice between April and October.

BOTANICAL NAME: *Acacia greggii*

COMMON NAMES: Catclaw or cat claw, paradise flower, devil's claw, wait-a-minute bush, *uña de gato, tésota, gatuño, palo chino* (Chinese stick), *tepame, algarroba,* tear blanket.

PROVENANCE: Southern California, Arizona, Southern Texas, and Mexico.

TERROIR: Prefers poor, dry soil in desert arroyos. Likes full sun in desert conditions. Rainy seasons cause the blooms to drop.

HONEY COLOR: Water white to medium amber.

TASTING NOTES: Rich aroma. Sweet, mild, and buttery flavor with a hint of iron. Heavy body. Granulates smoothly and with a waxy grain.

PAIRINGS: Drizzle over ricotta cheese and sliced fresh cantaloupe with prosciutto. Used in crafted-ale brewing.

16. CHESTNUT

PLANT CHARACTERISTICS: A tree or shrub valued for its large edible nuts and for its wood. The flowers are burrish and yellowish green and give way to two or three egg-shaped nuts in late autumn.

BLOOMS: May to July.

BOTANICAL NAME: *Castanea sativa* (*Castanea* from the name of the town of Kastania in Thessaly, Greece.)

COMMON NAMES: Spanish chestnut, American chestnut, chinquapin, *castagno, chât aignier, castaño*.

PROVENANCE: Native to southern Europe, growing in Turkey, Portugal, France (particularly in Corsica), Hungary, Italy, Croatia, Spain, and Bosnia.

TERRIOR: Requires a mild climate and adequate moisture for good growth. Is sensitive to late spring and early autumn frosts, and is intolerant of lime.

HONEY COLOR: Extremely dark amber, with a hue similar to chocolate. Sometimes reddish.

TASTING NOTES: Aromatic, pungent, heavy, rich, robust. Distinct nutty flavors with hints of wood, tar, and dried fruit. Sharp and bitter aftertaste.

PAIRINGS: Drizzle over Gorgonzola or stinky blue cheeses and walnuts, and serve with cabernet sauvignon. Pour over poached pears, and chocolate mousse.

17. CLOVE

PLANT CHARACTERISTICS: An evergreen with clusters of white or crimson flowers that produce an aromatic bud when dried. Beekeeping in Tanzania is very primitive. Beekeepers, primarily women, harvest honey from wild bees living in trees. Many keep bees in hollow coconut logs or gourd hives, and management of colonies is nonexistent. Honey is removed with heavy smoke or by using fire, which often destroys the colonies. Barely filtered, the honey is then exported to Oman, Jordon, where it is in high demand.

BLOOMS: Year-round.

BOTANICAL NAME: *Syzygium aromaticum,* syn. *Eugenia aromaticum* or *Eugenia caryophyllata*

COMMON NAMES: *Girpflier, marashi ya karafuu* (local name).

PROVENANCE: Pemba Island, Tanzania (a.k.a., the Spice Island).

TERRIOR: Sandy, maritime, ultratropical climate with high rainfall.

HONEY COLOR: Medium to dark amber.

TASTING NOTES: Highly aromatic. Exotic, spicy flavor that warms on the tongue.

PAIRINGS: Drizzle over fresh arugula, orange, and ginger salads with feta cheese, and serve with pinot noir. Spread on cinnamon toast or spice breads. Used to make wine and honey beer.

18. CLOVER

PLANT CHARACTERISTICS: The most diverse honey plant in the world; clover honey can be found in just about any pantry. Clover has three leaves; stems with four leaves are rare and considered lucky. The small flowers have round, spiky heads. Red clover has a large head of rose-purple flowers. Honeybees do not have a tongue long enough to pollinate these plants, but bumblebees do.

GENUS: *Trifolium or trefoil*

COMMON NAMES: White Dutch clover, alsike clover, red clover, *trifoglio, trèfle, trébol.*

PROVENANCE: Native of Europe and northern regions of the United States and Canada.

TERRIOR: Clover can be found along roadsides and railroads and in meadows. Prefers moist, temperate climates with lime soils and sufficient rain.

HONEY COLOR: Water white to extra light amber.

TASTING NOTES: Sweet, flowery, delicate flavor. Notes of green grass with a spicy aftertaste. Pleasant waxy texture. White clover honey is the honey all other honeys are compared to because it is thought to be the most pleasing to the widest variety of consumers.

PAIRINGS: Used to make glazes for ham. Delicious enough to eat by the spoonful.

19. COFFEE

PLANT CHARACTERISTICS: An evergreen shrub or small tree, Arabian coffee is a very important plant in commerce, providing the world with the most popular drink. Showy white blossoms smell like jasmine and are a valuable nectar and pollen source. The pollen is heavy and sticky. Blossoms give way to green berries that turn bright red and contain the beans that we roast to make coffee.

BLOOMS: All times of the year.

BOTANICAL NAME: *Coffea arabica*

COMMON NAMES: Coffee tree or shrub, mountain coffee.

PROVENANCE: Native to western and southern plantations of Ethiopia, Yemen, Vietnam, Brazil, and Thailand.

TERRIOR: The nectar flow is very intense after rains. Tolerates light shade and low temperatures, but not frost.

HONEY COLOR: Ranges from light to dark amber to brown and black.

TASTING NOTES: Delicate, rich, with notes of toffee, chocolate, brown sugar, and fragrant jasmine.

PAIRINGS: Drizzle over Parmesan cheese and chestnuts, and serve with merlot. Pour over dark chocolate, vanilla bean, or raspberry ice cream.

20. CORBEZZOLO

PLANT CHARACTERISTICS: An evergreen shrub with white, bell-shaped flowers. The edible fruit is round, prickly, and bright red and, despite the plant's common name, is not the same fruit as a strawberry.

BLOOMS: November to February.

BOTANICAL NAME: *Arbutus unedo*

COMMON NAMES: Strawberry tree, bitter honey (*miele amaro*).

PROVENANCE: Native to the Mediterranean region, especially Sardinia.

TERRIOR: Grows in lime soils. Prefers dry, hot summers.

HONEY COLOR: Medium to dark amber. Golden.

TASTING NOTES: Strong, intense flavor of raw nuts. Astringent, bitter, pungent, herbaceous, and slightly smoky.

PAIRINGS: Drizzle over pasta tossed with fresh ricotta cheese and freshly ground black pepper. Pour over gelato or prosciutto and melon. Used for making liquors and dessert wines.

21. CRANBERRY

PLANT CHARACTERISTICS: A semievergreen shrub producing dark pink flowers that resemble the neck and head of a crane—hence the original name *craneberry*. The fruits are shiny scarlet berries. Requires pollination by honeybees.

BLOOMS: May to June.

BOTANICAL NAME: *Vaccinium macrocarpon*

COMMON NAMES: American cranberry, bounceberry, large cranberry.

PROVENANCE: In the United States: Cape Cod, Massachusetts; Wisconsin; New Jersey; Oregon; and Washington. In Canada: Quebec.

TERRIOR: Acidic, sandy bogs in cooler regions of the northern hemisphere. Requires full sun.

HONEY COLOR: Medium amber with a rich reddish tint.

TASTING NOTES: Bright, fruity taste with strong, pungent hints of tart berries and deep, tangy plum notes. Crystallizes very quickly. This honey is high in vitamin C.

PAIRINGS: Drizzle over brie and walnuts, and serve with zinfandel. A Thanksgiving favorite spread for turkey. Mix into apple butter and vinaigrettes. Enjoy with dark chocolate and over spice breads.

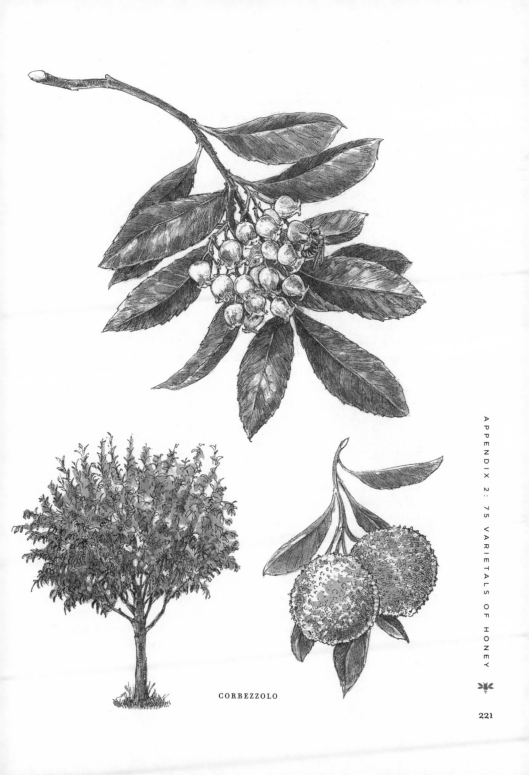

CORBEZZOLO

22. DANDELION

PLANT CHARACTERISTICS: Known to most people as a weed, this perennial is an important plant for honeybees, producing an abundance of pollen and nectar. The flower, a single, golden yellow blossom, is known as a wishie. When the seeds mature they form snowy white puffballs that are blown off by the wind.

BLOOMS: One of the first spring nectar sources. Blooms again in the fall.

BOTANICAL NAME: *Taraxcum officinale*

COMMON NAMES: Lion's tooth, blowball, yellow gowan, swine's snout, telltime, priest's crown, bitterwort, wild endive, Irish daisy, *tarassaco, piscialletto, dent de lion, diente de león*. (The English name *dandelion* is a corruption of the French *dent de lion*, meaning "lion's tooth," referring to the coarsely toothed leaves.)

PROVENANCE: Found worldwide. Native to Europe and Asia.

TERRIOR: Northern temperate regions. Grows in pastures, meadows, and wastelands.

HONEY COLOR: Deep golden yellow.

TASTING NOTES: Rich body. Strong, bitter, sharp flavor. Hints of chamomile and white pepper. Granulates quickly.

PAIRINGS: Drizzle over goat cheeses, figs, and pine nuts, and serve with sauvignon blanc. Used to make lemon vinaigrette, herbal jelly, and wine.

23. DZIDZILCHÉ

PLANT CHARACTERISTICS: A shrub or small tree with flowers that are fragrant and pale greenish yellow. An important source of nectar for bees.

BLOOMS: March to May.

BOTANICAL NAME: *Gymnopodium antigonoides* var. *floribundum*

COMMON NAMES: *Canelita, aguanales, tzitzilche, tzil tzil che, tsitsilché, nangaña.*

PROVENANCE: Yucatan, Mexico.

TERRIOR: Tropical forest with high humidity, intense rain, and dry seasons.

HONEY COLOR: Light amber.

TASTING NOTES: A high-quality, very fragrant honey with a unique flavor and aroma. Deep and tropical character, simultaneously wild and delicate. Must be used fairly quickly to avoid possible fermentation.

PAIRINGS: This honey is rarely exported and has traditionally been used for healing purposes.

24. EUCALYPTUS

PLANT CHARACTERISTICS: An aromatic tree or shrub with distinctively fragrant flowers that can be cream, pink, or red. The leaves and bark are covered with a blue-gray, waxy bloom that gives this plant the common name *blue gum*. Known as the fastest growing tree in the world.

BLOOMS: Beginning in spring and flowers all season.

BOTANICAL NAME: *Eucalyptus globulus*

COMMON NAMES: Tasmanian blue gum, *mallee, eucalipt, Eukaliptus.*

PROVENANCE: Native to Australia, Tasmania, and New Guinea. Grown in Arizona, Texas, California, and the Mediterranean.

TERRIOR: Temperate rain forest regions. Requires full sun.

HONEY COLOR: Pale yellow to amber, and sometimes dark, depending upon region.

TASTING NOTES: Strong woody aroma. Cool menthol and balsam flavors. Fruity and malty aftertaste. Thin bodied.

PAIRINGS: Drizzle over parmigiano reggiano, and serve with Chianti. Blends well with orange and lemon to make salad dressings and glazes or marinades for poultry. Used to mix martinis.

25. GREEK FIR

PLANT CHARACTERISTICS: An evergreen tree with dark green, bluish, needlelike leaves. The needles grow in a spiral, giving the tree an almost perfect pyramid shape. The cones are brown and resinous. The honey is actually made from the resin secreted in July and August, which technically makes it a honeydew.

BLOOMS: Produced in July and August. (Secreted by insects rather than blooms.)

BOTANICAL NAME: *Abies cephalonica*

COMMON NAMES: Christmas tree, fir tree, dwarf Greek fir, *melata.*

PROVENANCE: Native to Greece's Peloponnesos Mountains and Kefalinia Island. Also found in France, Austria, and Italy.

TERRIOR: Calcareous, well-drained soils along high mountains overlooking the sea. Prefers long periods of full sun.

HONEY COLOR: Cloudy, dark brown amber, red or green.

TASTING NOTES: Smoky, delicate flavor with resin, balsam, malt, caramel, menthol tones. Slow to granulate.

PAIRINGS: Drizzle over ricotta or goat cheeses and pine nuts, and serve with pinot noir or cabernet sauvignon. Mix with lemon as a marinade for lamb. Drizzle over Greek yogurt.

26. FIREWEED

PLANT CHARACTERISTICS: A tall perennial herb with reddish purple flowers that grow in large, spiky clusters. Pollen is deep green-blue to purple. After a forest fire, fireweed is one of the first plants to grow on the burnt fields.

BLOOMS: July to late September.

BOTANICAL NAME: *Epilobium angustifolium*

COMMON NAMES: Great willow herb, evening primrose, rosebay, Indian pink, *epilobio, épilobe, epilobi.*

PROVENANCE: In the United States: Washington, Oregon's Cascade Mountains, southern Michigan, Alaska, and Minnesota. Also found in Canada.

TERROIR: Thrives in moist, clay-rich humus, yet tolerates sandy or rocky soils and cool temperatures. Cool nights and warm days produce the best nectar flow.

HONEY COLOR: Absolutely water white.

TASTING NOTES: Known as the champagne of honeys. Delicate, fruity, and smooth. Subtle, tealike notes with hints of spice and butter. Granulates quickly.

PAIRINGS: Drizzle over Brie topped with chopped fresh cranberries on a baguette, serve with champagne or prosecco. Mix with butter to spread over pound cakes.

27. GALLBERRY

PLANT CHARACTERISTICS: A tall evergreen shrub with white flowers that produce so much nectar it can be seen shining on the leaves. Black berries hang on to the tree throughout the winter. An important honey plant in Georgia.

BLOOMS: May to June.

BOTANICAL NAME: *Ilex glabra*

COMMON NAMES: Inkberry, evergreen winterberry, Appalachian tea, dye-leaves.

PROVENANCE: Georgia and Florida.

TERROIR: Moist to wet, swampy, sandy, acidic soil found along coastal regions. Ideal production, according to beekeepers, occurs when the gallberry bush has "feet in water, head in sunshine."

HONEY COLOR: Light amber with a yellowish cast.

TASTING NOTES: Citrusy, spicy, tangy, minty flavor and a slight burning aftertaste. Aromatic. Thick and heavy body. Does not granulate even in cold temperatures.

PAIRINGS: Drizzle over sliced fresh melon, feta cheese, and chopped fresh mint, serve with sauvignon blanc.

28. GOLDENROD

PLANT CHARACTERISTICS: A perennial with long beautiful branches and clusters of bushy golden flowers that produce large quantities of nectar. Goldenrod is an important nectar source for honeybees and must be cross-pollinated by insects and honeybees.

BLOOMS: August through October. Many people think they are allergic to goldenrod, but in fact it is ragweed that is usually responsible for causing reactions.

BOTANICAL NAME: *Solidago rigida*

COMMON NAMES: Rigid goldenrod, solidago, *verge d'or, solidage, vara de oro.*

PROVENANCE: Native to North America, especially New England, eastern Canada, and Mexico; also found in Hawaii.

TERROIR: Tolerates poor, dry soil. Thrives in full sun or shade. Found in meadows, fields, and open woods, and along highways and train tracks.

HONEY COLOR: Brilliant amber to a warm golden yellow.

TASTING NOTES: Bright, sharp, floral-like flavor with hints of fresh straw. Spicy aftertaste. Granulates coarsely.

PAIRINGS: Drizzle over brie or goat cheeses and toasted pine nuts, serve with chardonnay or canaiolo. Spread on cinnamon raisin toast or challah bread.

29. HEATHER, BELL

PLANT CHARACTERISTICS: A low-growing ornamental shrub with flowers that are white to pink and also include a wide range of purples and reds. Bell heather is known throughout the Scottish Highlands.

BLOOMS: Late July to November.

BOTANICAL NAME: *Erica cinerea*

COMMON NAMES: Heather bell, erica, *erica cinerea, bruyère cendrée, brezo ceniciento.*

PROVENANCE: Scotland and France.

TERRIOR: Drought tolerant. Prefers well-drained soil and full sun.

HONEY COLOR: Very dark amber with a tinge of red.

TASTING NOTES: One of the most well-respected honeys of the world. A strong, dark honey. Slightly bitter with an earthy aftertaste of burnt caramel, fudge, licorice, rum, and plums. Does not crystallize.

PAIRINGS: Drizzle over Stilton or cheddar cheeses, and serve with cabernet sauvignon. The Scottish Highlands' famous heather-honey cake is made with bell heather honey.

30. HEATHER, LING

PLANT CHARACTERISTICS: A low-growing perennial shrub with evergreen leaves and purple stems with pink-lilac flowers that produce brown to golden red-brown pollen. Ling heather honey is highly respected, commanding high prices.

BLOOMS: August to September. Beehives are moved to the moorland and heathland regions each summer.

BOTANICAL NAME: *Calluna vulgaris* (*Calluna* is derived from a Greek word meaning "to sweep," and the plant was used to make brooms.)

COMMON NAME: Ling heather, spring torch, *brande, bruyère commune, erica vulgaris, callune, brezo.*

PROVENANCE: Scottish Highlands.

TERROIR: The dominant plant of the Highlands, it covers moors, heaths, and open woods, thriving in the damp, rocky, acidic soil, and in both sunny and shaded locations.

HONEY COLOR: Dark amber with a reddish hue.

TASTING NOTES: Aromatic. Strong, slightly bitter, smoky, toffeeish flavor with plum tones. Known as "the port wine of honeys," ling heather honey has an unusually thick, jellylike consistency. It is too thick to extract from the comb, so it must be gently pressed out. The honey can be stirred until it becomes liquid, but will return back to its original jellylike state. Does not crystallize.

PAIRINGS: Drizzle over Stilton or cheddar cheeses, and serve with cabernet sauvignon. Use in relish and chutney. Makes a nice honey beer.

31. HORSEMINT

PLANT CHARACTERISTICS: A perennial herb with lance-shaped leaves and tubular yellow flowers with purple tinges and spots. Horsemint is an important honey plant, attracting butterflies, hummingbirds, and honeybees.

BLOOMS: June to October.

BOTANICAL NAME: *Monarda punctata*

COMMON NAMES: Spotted horsemint, spotted bee balm, dotted mint.

PROVENANCE: Native to the United States. Most important honey plant in the Black and Grand prairies of Texas.

TERROIR: Sandy prairies, open fields, roadsides, and hills. Prefers lime soil and full sun.

HONEY COLOR: Clear water white to amber.

TASTING NOTES: Highly aromatic, spicy, and minty in flavor. Hints of lemon and oregano.

PAIRINGS: Serve with Havarti or Asiago cheeses and merlot. Stir into herbal teas and mix into mint juleps. Use in glazes for grilling meats.

HORSE MINT

32. HUAJILLO (pronounced *wa-He-yo*)

PLANT CHARACTERISTICS: A small-to-medium-sized desert bush not to be confused with the Guajillo chili pepper. The blooms are fragrant, small, round, and creamy white to yellow in color, with globular heads.

BLOOMS: March to early May.

BOTANICAL NAME: *Acacia berlandieri*

COMMON NAMES: Berlandier acacia, guajillo acacia, guajilla or huajilla.

PROVENANCE: Uvalde County in southwestern Texas chaparral or brush country. Unique to the Edwards Plateau of Texas and to southwestern Texas and northern Mexico

TERRIOR: Sandy, rocky soil in dry desert and uplands. Prefers intense sun.

HONEY COLOR: Mild, light-colored, crystal white with a pearly reflection, like new milk, or a very light amber.

TASTING NOTES: Extremely delicate and distinctive taste that is described as very light, smooth, and tangy. Hints of apricot, prune, and cranberry.

PAIRINGS: Drizzle over Brie and macadamia nuts, and serve with Sauternes, port, or Madeira. Use in pineapple-tamarind glaze for pork chops and lamb.

33. KAMAHI (pronounced *car-MY*)

PLANT CHARACTERISTICS: A slow-growing evergreen tree with small, creamy white flowers in erect spikes, producing white pollen.

BLOOMS: October to December (New Zealand spring).

BOTANICAL NAME: *Weinmannia racemosa*

COMMON NAMES: Red birch.

PROVENANCE: Produced in both the North and South islands of New Zealand.

TERRIOR: Rugged native forests or shrublands. Prefers full sun and a temperate climate.

HONEY COLOR: Golden yellow to light amber to green.

TASTING NOTES: Strong-flavored honey with hints of musky toffee and molasses, balanced by bitter flavors. Preferred by many honey connoisseurs. Full-bodied and finely crystallized texture.

PAIRINGS: Drizzle over blue or Brie cheeses and serve with Syrah. Spread on to pain d'epices (spice bread) and biscuits. Use as a glaze for fish and meats.

35. KIAWE (pronounced *kee-AH-vay*)

PLANT CHARACTERISTICS: An invasive thorny bush in the same family as mesquite, with small flowers in pale yellow spikes that bear light yellow bean pods. Kiawe is an excellent honey tree.

BLOOMS: When three to four years old, frequently flowers twice a year in March and September.

BOTANICAL NAME: *Prosopis pallida*

COMMON NAMES: Huarango, American carob, algarroba, bayahonda.

PROVENANCE: Native to Colombia, Ecuador, and Peru, Naturalized in Hawaii; the honey comes from only the Puako Forest of Hawaii's big Island, where it has been harvested for one hundred years.

TERROIR: Grows in the arid, tropical volcanic lava of an isolated oasis in the coastal forest.

HONEY COLOR: Pearly water white with waxy, golden overtones.

TASTING NOTES: Rich tropical fruit and menthol flavors. A smooth, creamy consistency and extremely fine crystals that produce an effervescent sensation on the tongue. This honey crystallizes very, very rapidly, right in the comb, which gives it the lovely texture.

PAIRINGS: Drizzle over Asiago or Havarti cheeses on flatbread crackers topped with sliced fresh kiwi and cashews, and serve with champagne or chardonnay.

36. KNAPWEED

PLANT CHARACTERISTICS: A herbaceous, bushy, branched weed with rough green leaves that have spots, giving it the name *spotted knapweed*. The flowers are white or purple-to-pink and have fringy petals that resembles cornflowers. An excellent nectar source for honeybees.

BLOOMS: Early to late June to August.

BOTANICAL NAME: *Centaurea maculosa*

COMMON NAMES: European star thistle, ballast-waif, spotted knapweed.

PROVENANCE: Native to eastern Europe. Grows in Idaho, Alaska, Georgia, Mississippi, Oklahoma, Texas, Michigan, New York, and Montana.

TERROIR: Requires full sun. Grows in dry, rocky prairies and along railroad tracks, waterways, and grasslands.

HONEY COLOR: Light to medium amber.

TASTING NOTES: Bitter, tangy, and astringent flavors. Full-bodied and thick. Slow to granulate.

PAIRINGS: Best used for mead making.

37. KNOTWEED

PLANT CHARACTERISTICS: A herbaceous perennial plant. The hollow stems with raised nodes resemble bamboo. Small, creamy white flowers are an important source of nectar in late summer when little else is in bloom. Knotweed honey is sometimes referred to as bamboo honey.

BLOOMS: August to October.

BOTANICAL NAME: *Fallopia japonica*, syn. *Polygonum cuspidatum, Reynoutria japonica*

COMMON NAMES: Fleeceflower, monkeyweed, Huzhang Hancock's curse, elephant ears, pea shooters, donkey rhubarb (although it is not a rhubarb), Japanese bamboo.

PROVENANCE: Native to eastern Asia (Japan, China, and Korea), the northeastern United States (Pennsylvania and New York), and Europe.

TERRIOR: Moist soil, wetlands.

HONEY COLOR: Very dark amber with reddish tones.

TASTING NOTES: Rich, heavy, and robust. Hints of caramel, brown sugar, and maple flavors. A fruitier version of buckwheat honey without the malty flavors.

PAIRINGS: Drizzle over Camembert, fresh figs, and pecans, and serve with merlot. Spread over waffles, pancakes, gingerbreads, banana muffins, and rum cakes. Mix into custards, tapiocas, and rice puddings.

38. KUDZU

PLANT CHARACTERISTICS: A climbing, woody, perennial vine that is highly invasive and weedy. Spiky, purple-violet flower with highly fragrant aromas. Copious nectar producer.

BLOOMS: Late summer.

BOTANICAL NAME: *Pueraria thunbergiana*

COMMON NAMES: Known as foot-a-night vine, mile-a-minute vine, porch vine, telephone vine, and wonder vine because kudzu can grow aggressively. *Gé Gan* in traditional Chinese medicine.

PROVENANCE: Native to China and Japan. Found in Alabama's Choccolocco Valley, as well as in Georgia and Mississippi.

TERROIR: Deep, loamy soils.

HONEY COLOR: Bluish to very dark purple.

TASTING NOTES: A rare honey with a strong flavor. Fruity notes of apple, peach, grape, and bubblegum.

PAIRINGS: Drizzle over ricotta cheese and sliced green apples, and serve with Riesling or cabernet franc. Add to glazes for pork tenderloin. Mix into lemonade garnished with sprigs of fresh mint. Used to make famous kudzu blossom jelly.

39. LAVENDER

PLANT CHARACTERISTICS: A small shrubby camphorous plant. Aromatic plant with a camphor scent; slender leaves; deep bluish, thin stalks; and purple flowers. Lavender honey is highly respected in France and commands premium prices.

BLOOMS: From July to August.

BOTANICAL NAME: *Lavendula angustifolia*

COMMON NAMES: Lavender, *lavanda, lavendel, espliego*.

PROVENANCE: Native to Southern Europe, especially Provence, France; Catalonia, Spain; and Portugal.

TERROIR: Thrives in sunny exposures and dry, poor, acidic soils.

HONEY COLOR: Light amber.

TASTING NOTES: Strong camphor-perfumed taste with sweet tobacco notes. Smooth, rich body.

PAIRINGS: Drizzle over blue and goat cheeses and almonds, and serve with cabernet sauvignon. Pour over vanilla bean ice cream. Mix into salad dressings with lemon zest to create a glaze for grilled chicken breasts. Stir into chamomile tea.

40. LEATHERWOOD

PLANT CHARACTERISTICS: An evergreen tree and one of the largest flowering trees; attains maturity at around 250 years old. So named because of the leathery texture of its leaves. Delicate white flowers with a spicy scent. Leatherwood honey of Tasmania has a reputation throughout the world.

BLOOMS: Mid-January until early March (Tasmanian summer).

BOTANICAL NAME: *Eucryphia lucida*

COMMON NAME: Swamp gum, mountain ash.

PROVENANCE: Tasmania, Australia.

TERRIOR: The wetter forest regions throughout the western portion of Tasmania. The leatherwood tree is part of Tasmania's World Heritage Zone of protected forest, a botanically unique area of rain forest and acidic soils.

HONEY COLOR: Golden to medium amber.

TASTING NOTES: Rich, subtly sweet, spicy, slightly acidic. Subdued but persistent rose, violet, and caraway notes.

PAIRINGS: Drizzle over Tasmanian smoked cheddar or Australian cheeses, serve with sourdough bread and cabernet sauvignon. Used to make ginger beers and ales.

41. LEHUA (pronounced *Lay-WHO-uh*)

PLANT CHARACTERISTICS: Common evergreen tree with fiery red lehua flowers, which are very popular with bees. The flowers are the symbols of erotic love and are sacred to Pele, the goddess of fire and volcanoes, and to Laka, the goddess of hula. The honey is rare and highly sought after.

BLOOMS: May to July.

BOTANICAL NAME: *Metrosideros polymorpha*

COMMON NAME: Ohia lehua.

PROVENANCE: Native to the forests of the volcano Mauna Loa in the remote Ka'u district on the big island of Hawaii.

TERROIR: Found in rain forest and rolling pastures on the north side of the volcanoes. Requires full sun.

HONEY COLOR: Water white to light golden amber.

TASTING NOTES: Fragrant and delicate with a buttery flavor. Overtones of butterscotch, English toffee, and lilies. Lehua crystallizes extremely fine, making it naturally creamy and spreadable.

PAIRINGS: Drizzle over blue cheese and sliced fresh pears, and serve with Sauternes. Stir into green tea.

42. LEMON

PLANT CHARACTERISTICS: The edible fruit of the citrus-evergreen tree or bush. Its white, fragrant flowers have purple edges. Lemon honey is typically harvested in the Mediterranean. The United States markets lemon honey as orange honey since they are most often blended.

BLOOMS: Throughout the year.

BOTANICAL NAME: *Citrus limon*

COMMON NAMES: Lemon, citrus, *limone, citron, limón*.

PROVENANCE: Native to India, Burma, and China. Later introduced to Persia and then to Iraq and Egypt. Grows today in Italy, Spain, Argentina, Israel, and in the United States, especially California, Arizona, and Florida.

TERROIR: Originating in tropical and subtropical climates, but frost hardy. Sunny, humid environment with fertile soil.

HONEY COLOR: Bright amber yellow.

TASTING NOTES: Strongly scented, citrusy, bright, tart, sour, aromatic. Good source of vitamin C.

PAIRINGS: Drizzle over ricotta or Taleggio cheeses, and serve with chardonnay. Spread over poppy-seed breads. Blends well with thyme, garlic, basil, and ginger dressings. Used to make limoncello.

43. PURPLE LOOSESTRIFE

PLANT CHARACTERISTICS: Invasive wetland plant with dense, spiky, purple or pink flowers producing a yellow to green or violet pollen. A favorite pollen and nectar source for honeybees.

BLOOMS: July to October.

BOTANICAL NAME: *Lythrum salicaria*

COMMON NAMES: Rebel weed, spiked loosestrife, spiked lythrum, *riparella, salicaire, salicaria, qian qu cai.*

PROVENANCE: Native to Eurasia. Found in the United States along the northern Atlantic coast and in the Midwest.

TERROIR: Shallow freshwater pond edges, marshes, and bogs. Fertile, neutral to slightly alkaline soils.

HONEY COLOR: Extremely dark purple; resembles motor oil.

TASTING NOTES: Rich, strong, unappealing. It has a delicately sweet flavor and is said to be highly aromatic. Heavy body.

PAIRINGS: Strictly bakery-grade honey used for cookies and spreads. Good choice for making wine or mead.

44. MACADAMIA

PLANT CHARACTERISTICS: The name refers to the exotic and rare round edible nut that is highly nutritious. This evergreen tree has creamy white, pink, or purple flowers that grow in long clusters.

BLOOMS: February to April (Australian summer).

BOTANICAL NAME: *Macadamia integrifolia*

COMMON NAMES: Queensland nut, bush nut, maroochi nut, bauple nut; indigenous Australian names include *gyndl, jindilli,* and *boombera.*

PROVENANCE: Native to New South Wales and Queensland in Australia. Found in the United States in Hawaii, Arizona, and San Diego, California.

TERROIR: Plantations in tropical, mild, high-rainfall, and frost-free climates. Fertile, well-drained soils.

HONEY COLOR: Medium amber to dark, deep amber.

TASTING NOTES: Exotic, rich, not too sweet, with delicious tangy, musky floral undertones. Hints of velvety butterscotch and nuts.

PAIRINGS: Drizzle over ricotta or goat cheese with sliced fresh pineapples or passion fruits, and serve with sauvignon blanc or zinfandel. Enjoy with dark or white chocolate and coconut desserts. Drizzle over banana cakes, vanilla ice cream, fruit salads, and waffles. Used to make honey butter.

45. MAUKA

PLANT CHARACTERISTICS: An evergreen bush with prickly leaves dotted with oil glands. When bruised, these leaves give off a gingery, peppery smell. Its white or pink fragrant flowers have white pollen.

BLOOMS: September to June (Australian summer).

BOTANICAL NAME: *Leptospermum scoparium*

COMMON NAMES: Tea tree, New Zealand honeysuckle, broom tea tree, *manuka* in the local Miori language.

PROVENANCE: New Zealand and southern Australia.

TERRIOR: Found in sandy coastal woodlands with well-drained, infertile acidic soil and full sun.

HONEY COLOR: Medium to dark amber with an orange tint.

TASTING NOTES: Intense medicinal flavors with overtones of burnt sugar. Also gingery, peppery, earthy, woody flavors. Heavy body. Often sold crystallized.

PAIRINGS: Manuka honey is known around the world for its healing properties and is taken mostly as a natural-health remedy.

46. MANZANITA

PLANT CHARACTERISTICS: An evergreen shrub with clusters of bell-shaped white flowers and red-orange bark. The edible berries look like tiny apples and are white when new, then turn red-brown as the summer progresses. Popular nectar source for butterflies, hummingbirds, and honeybees.

BLOOMS: Early in the year, sometimes as early as January.

BOTANICAL NAME: *Arctostaphylos manzanita*

COMMON NAMES: Common manzanita, whiteleaf manzanita. (The word *manzanita* is the Spanish diminutive of *manzana,* meaning apple. A literal translation would be "little apple.")

PROVENANCE: Native to California, where it can be found in the Coast Ranges and Sierra Nevada foothills.

TERROIR: Chaparral slopes and low-elevation coniferous forest ecosystem. Cool nights stimulate nectar flow. Full sun. Clay or sandy soil.

HONEY COLOR: Dark, golden amber.

TASTING NOTES: Full-bodied. Intense floral notes. Smoky, woody, rich, and earthy with a slightly peppery and tangy flavor. Granulates quickly.

PAIRINGS: Drizzle over goat cheese and fresh apricots, and serve with pinot noir or prosecco. Used to make ciders and meads.

47. MESQUITE

PLANT CHARACTERISTICS: A small deciduous shrub or scraggly tree with droopy branches of feathery foliage bearing spikes of yellow beans or pods. Long clusters of fragrant yellow-orange flowers.

BLOOMS: April and again in June.

BOTANICAL NAME: *Prosopis glandulosa*

COMMON NAMES: Honey mesquite, honey pod, *haas*. Called *mizquitl* in Nahuatl, the Aztec language.

PROVENANCE: Native only to California, Arizona, New Mexico, Texas, and the Chihuahuan Desert of Mexico.

TERROIR: Shallow, loamy, sandy, desert prairies. Prefers well-drained soil and full sun. Tolerates heat and is drought resistant. Abundant rain and then hot periods are ideal for nectar flow.

HONEY COLOR: Medium amber with a brown tint.

TASTING NOTES: Sweet, yet warm, smoky, woody, citrus flavors. Light, delicate, and aromatic. Granulates quickly.

PAIRINGS: Drizzle over goat cheese with sliced fresh mango, and serve with champagne or Sauternes. Mix with lime for barbecue sauces and rubs for smoked ham or ribs. Spread on blue-corn pancakes, whole-grain breads, and corn muffins.

48. MILKWEED

PLANT CHARACTERISTICS: A perennial herb with global clusters of purplish to pink flowers that are an important nectar source for bees. The flowers are called pinch-trap flowers because insects often get their feet or antennae stuck in the slit where the pollen is found. When the honeybee leaves, it often carries away parts of the flower in addition to the pollen.

BLOOMS: July through August.

BOTANICAL NAME: *Asclepias syriaca* (*Asclepias* comes from the name of the Greek god of medicine, *Asklepios*.)

COMMON NAMES: Butterfly flower, silky swallow-wort, silkweed. (The name *milkweed* refers to the milky white sap that seeps out of the stems and lance-shaped leaves when they are broken.)

PROVENANCE: Native to Oklahoma and the southern peninsula of Michigan.

TERROIR: Full sun. Prefers moist loam or clay sand prairies, but tolerates a variety of soils.

HONEY COLOR: Water white with a light yellow tinge.

TASTING NOTES: Very heavy in body, fruity, quincelike, slight spicy tang. Slow to granulate

PAIRINGS: Adds a touch of spice to salad dressings and vegetables.

49. ORANGE BLOSSOM

PLANT CHARACTERISTICS: The small flowering tree, known as sweet orange, bears the citrus fruit that is actually considered a berry. This small evergreen tree has oval leaves and white fragrant blossoms. The orange fruit is highly valued for its oils. Honey harvested from grapefruit, lemon, tangelo, tangerine, and orange blossoms are usually marketed in the United States as simply orange-blossom honey, with no distinction between the sources.

BLOOMS: March to April.

BOTANICAL NAME: *Citrus sinensis*

COMMON NAMES: Orange blossom, *arancia, naranjo*.

PROVENANCE: Native to Asia. Citrus groves flourish in Spain, Mexico, and Israel, and in the United States in California, Florida, Arizona, and parts of Texas around the Gulf of Mexico.

TERROIR: Tropical to subtropical climate; regions free of frost. Light, loamy, moist soils. Secretes abundant nectar when the climate is very warm with no fog.

HONEY COLOR: Light to medium amber with bright orange tint.

TASTING NOTES: Classical citrus aroma. Acidic, distinctive fruity orange taste. Hints of rose and jasmine, as well as beeswax.

PAIRINGS: Drizzle over goat cheese and toasted pine nuts, and serve with sauvignon blanc. Use in glazes for pork chops, ham, or chicken wings. Ideal for use in marmalades, cranberry sauces, pico de gallo, and frosting for carrot cakes.

50. TURKISH PINE

PLANT CHARACTERISTICS: This fast-growing evergreen tree is host to a sap-sucking aphid, *Marchalina hellenic,* that secretes a sugar called honeydew. Honeybees collect the honeydew and turn it into pine honey. The honey contains valuable minerals and vitamins.

BLOOMS: Produces cones that open slowly over a year or two to release its seeds.

BOTANICAL NAME: *Pinus brutia*

COMMON NAMES: Calabrian pine, east Mediterranean pine, Brutia pine, and *çam bal*, which means "pine honey" in Turkish.

PROVENANCE: Aegean region of Turkey, Greece, Italy, France.

TERROIR: Prefers mild, wet winters and dry, very warm summers.

HONEY COLOR: Light amber.

TASTING NOTES: Sweet and spicy, with rich, woody pine flavors. Considered a valuable honey.

PAIRINGS: A popular Turkish breakfast is bread or yogurt topped with pine honey butter.

51. PITCAIRN ISLAND

A rare honey named after its place of origin. The Pitcairn Islands, in the South Pacific, were settled in 1790 by mutineers from the *HMS Bounty*. The island's bee population has been certified as disease free, and Pitcairn honey is one of the island's main economic resources. The island's approximately fifty residents and thirty beehives produce all of the honey. If you are inclined to hunt down this delicious honey, it will take three to five months to receive written permission from the government.

PROVENANCE: Pitcairn Island.

TERROIR: Humid, tropical climate with moderate rainfall and sandy fertile soil.

HONEY COLOR: Creamy white.

TASTING NOTES: Thick body. Smooth and cool on the tongue. Tropical flavors, including notes of mango, lata, passion fruit flower, guava, and rose apple flowers.

PAIRINGS: Decadent enough to eat straight from the spoon. Mix with ginger and cashews as a marinade for duck.

52. PRICKLY PEAR

PLANT CHARACTERISTICS: These evergreen segmented cacti spread in clumps and have yellow or red flowers, which grow on the green oval pads that look like a beaver tail. Prickly pear or *tuna* is the edible red fruit that grows on the cactus and is named for its yellow needles, called glochids.

BLOOMS: July till September.

BOTANICAL NAME: *Opuntia engelmanni*

COMMON NAMES: Indian fig, Cactus apple, cow's tongue cactus, desert prickly pear, Texas prickly pear, calico catus, Engelmann's prickly pear, *fico d'india, nopal, abrojo, joconostle,* and *vela de coyote.*

PROVENANCE: Texas; Baja; California; Sonoran Desert; Arizona; Chihuahua, Mexico: and Sicily

TERROIR: Arid desert regions with sandy soils with good drainage. Needs full sun and will survive on very little water once established.

HONEY COLOR: Medium to dark amber with bright red tint.

TASTING NOTES: Heavily bodied. Granulates in large crystals that float in the liquid rather than at the bottom. For this reason, this honey is called *buttermilk* honey. Fragrant, floral, musky, tangy, often bitter. Hints of watermelon, strawberries, and figs.

PAIRINGS: Drizzle over ricotta or Gorgonzola and serve with prosciutto and pinot grigio. Mix with lime and ginger as a glaze for pork chops or tenderloin.

53. PŌHUTUKAWA (pronounced *pō-hu-tu-ka-wa*)

PLANT CHARACTERISTICS: An evergreen tree that, at around Christmas time, explodes with masses of brilliant crimson to deep bloodred flowers. The Pōhutukawa tree lives to be a thousand years old yet blooms infrequently, making this honey rare.

BLOOMS: November to peaking in December (New Zealand winter).

BOTANICAL NAME: *Metrosideros excelsa*

COMMON NAMES: New Zealand Christmas tree or fire tree.

PROVENANCE: Coastal areas of the north and Three Kings Islands of New Zealand, as well as Rangitoto, an island in the Hauraki Gulf.

TERRIOR: Coastal forest on cliffs overhanging the sea. Light, sandy, moist soils. Sunny locations.

HONEY COLOR: Pale—the whitest honey in world.

TASTING NOTES: Known as the Queen's honey, this honey has a distinct butterscotch flavor with wonderful floral, citrusy lime flavors followed by a slightly salty finish.

PAIRINGS: Drizzle over Emmental and Gruyére or mild Havarti.

54. RASPBERRY

PLANT CHARACTERISTICS: The name refers to the edible sweet-tart fruit of the thorny American red raspberry shrub. Usually, the cane growth is attained the first year, then the fruit is produced the second year. Raspberry is an excellent honey plant.

BLOOMS: June through September.

BOTANICAL NAME: *Rubus strigosus*

COMMON NAMES: American red raspberry, *lampone, framboisier, frambueso*.

PROVENANCE: Native to North America. A leading honey plant in parts of Maine, New Hampshire, Vermont, Washington, and Wisconsin.

TERRIOR: Typically found in acidic, sandy forest soils. Prefers full sun.

HONEY COLOR: Extra light amber with reddish yellow hues.

TASTING NOTES: Smooth and floral, fruity yet tart, with a distinctive raspberry flavor. Crystallizes rapidly.

PAIRINGS: Drizzle over goat or Brie cheeses and walnuts, and serve with dessert wines or champagne. Stir into Earl Grey tea or lemonade. Use to make vinaigrettes and mint jams.

55. RĀTĀ

PLANT CHARACTERISTICS: Southern Rātā is a massive flowering tree with dark green leaves with indented tips and sprays of bright red flowers with dark stamens. Forests of flowering Rātā look like a red carpet covering the mountains on which it grows.

BLOOMS: January to March (New Zealand summer), sporadically every three to five years.

BOTANICAL NAME: *Metrosideros umbellata*

COMMON NAMES: Ironwood

PROVENANCE: West coast of New Zealand's South Island.

TERROIR: Wet coastal forest and lowlands with cooler temperatures. Needs full sun.

HONEY COLOR: Very light or white crystallized.

TASTING NOTES: Rātā honey is one of the finest honeys in the world. Silky and creamy with a rich floral flavor. Distinctively salty. One of the fastest crystallizing honeys.

PAIRINGS: Drizzle over pecorino and serve with sliced fresh pears. Drizzle over ice cream, muffins, and panna cotta. Makes a wonderful honey-mustard marinade for beef.

56. REWAREWA (pronounced *rewa-rewa*)

PLANT CHARACTERISTICS: An evergreen tree with distinctive clustered, brick red flowers that split and curl into spirals.

BLOOMS: October to December (New Zealand late spring).

BOTANICAL NAME: *Knightia excelsa*

COMMON NAMES: New Zealand honeysuckle or alpine flower.

PROVENANCE: Native to the bush areas and the Nelson/Marlborough area of New Zealand's North Island and the Marlborough Sounds at the top of the South Island.

TERROIR: Low elevation, valley forests, sea-drowned valleys.

HONEY COLOR: Burnished amber or reddish brown.

TASTING NOTES: Malty, smoky flavor with overtones of caramel, dried fruit, and ginger. This honey is robust and woody. Very thick bodied, yet buttery in texture. Rewarewa honey has significant levels of antioxidants.

PAIRINGS: Serve with flaky croissants, granola, or yogurt. Mix into marinades for strong, savory game.

RHODODENDRON

57. RHODODENDRON

PLANT CHARACTERISTICS: Bush with large, showy flowers that have five pink petals.

BLOOMS: Late June to July.

BOTANICAL NAME: *Rhododendron hirsutum*

COMMON NAMES: Hairy alpine rose, alpen rose.

PROVENANCE: Italian Alps, Austria, Switzerland, and France.

TERROIR: High altitudes, full sun, neutral to acidic soil.

HONEY COLOR: Light to medium amber.

TASTING NOTES: Medium sweetness, light aroma, and subtle flavors. Sharp, floral, and herbal flavors with undertones of wild berries, watermelon, and wet moss.

PAIRINGS: Drizzle over Val d'Aosta cheese and sliced fresh pears or figs, and serve with vin santo. Mix with cinnamon to season winter squash.

58. ROSEMARY

PLANT CHARACTERISTICS: A woody perennial with a distinctive camphor aroma. It has narrow evergreen leaves. The flowers are white, pink, purple, or blue and attractive to honeybees.

BLOOMS: April to June.

BOTANICAL NAME: *Rosmarinus officinalis*

COMMON NAMES: *Rosmarino, romarin, romero.*

PROVENANCE: Found in the Mediterranean regions of Spain, Italy, and France.

TERROIR: Dry, sandy, rocky soils. Full sun. Warm summers and dry winters.

HONEY COLOR: Clear water white with a tinge of straw.

TASTING NOTES: Fresh herbal and floral flavors. Hints of lemon and pine. Granulates quickly.

PAIRINGS: Drizzle over blue or Camembert cheeses and serve topped with walnuts on olive oil focaccia bread. Add to glazes for chicken and roasted potatoes. Mix with vodka and lemon for a martini.

59. SAGE

PLANT CHARACTERISTICS: A bushy shrub with aromatic leaves and white flowers that have a blue tinge. The foliage secretes a highly pungent, sticky oil. An attractive nectar plant for honeybees.

BLOOMS: April to July.

BOTANICAL NAME: *Salvia mellifera* (from the Latin word salveo, meaning "to save because of its medicinal value")

COMMON NAMES: Ball sage; black button sage; purple, black, and white sages; *salvia*; *sauge*.

PROVENANCE: Native to the Mediterranean and the California coast.

TERRIOR: Canyons, high hills, sunny slopes. Prefers full sun and low humidity.

HONEY COLOR: Water white, light amber with green tinge.

TASTING NOTES: Elegant with an herbal, pungent, slightly warm flavor. A bit of pepper and anise with a floral essence. Heavy bodied. Slow to granulate.

PAIRINGS: Drizzle over Parmesan or manchego cheeses on olive oil crackers, and serve with burgundy. Mix into lemonade and herbal teas. Use as a glaze for veal, pork, beef, and game.

60. SAINFOIN (pronounced *sān-fóin*)

PLANT CHARACTERISTICS: A perennial herb used for animal fodder. The flowers are snowy white or pale pink and form in long spikes. The pollen, which is saffron yellow, and nectar are attractive to honeybees and yield good amounts of honey.

BLOOMS: June to September.

BOTANICAL NAME: *Onobrychis viciifolia*

COMMON NAMES: Esparcet, holy hay, holy grass, holy clover, *crocette*. (It is named *holy clover* because baby Jesus is said to have laid his head in this grass in the manger where he slept.) *Sainfoin* means healthy hay in French.

PROVENANCE: Native to Eurasia.

TERRIOR: Thrives in chalky rock soils, limestone soil, grassland, cultivated land, and wastelands. Prefers warm climates.

HONEY COLOR: Pale straw yellow.

TASTING NOTES: Sweet, fragrant, and floral with a spicy, delicate flavor. Medium bodied with a grainy and creamy texture. Crystallizes quickly.

PAIRINGS: A delightful spreading honey for baguettes, scones, and toasted brioche.

61. SAW PALMETTO

PLANT CHARACTERISTICS: A tall tree or shrub with yellow-green foliage all year round. Yellow-white flowers with large reddish black fruit.

BLOOMS: April to May.

BOTANICAL NAME: *Serenoa repens*

COMMON NAMES: Saw palmetto

PROVENANCE: The southeastern United States, most commonly along the Atlantic and Gulf Coast plains of North Carolina, Texas, and Florida.

TERRIOR: Sand ridges, flatwood forests, coastal dunes, and islands near marshes.

HONEY COLOR: Light yellow to amber

TASTING NOTES: Salty and citrusy with anise, prune, and herbal, woody overtones. Thinner in body than most honeys.

PAIRINGS: Drizzle over hard cheeses and serve with ham or prosciutto. Mix as dressings for tart greens or citrus salads. Stir into black teas.

62. SIDR

PLANT CHARACTERISTICS: An ancient tree considered sacred to the Muslims. This climbing evergreen has yellow fruit that is edible when it turns red. The clustered flowers are pale yellowish green. Sidr honey of Hadramaut is the most expensive and prized honey in the world. Yemen's beekeepers ensure the purity of their honey and will allow their bees to die rather than feed them sugary syrup. The honey has an unusually high level of antioxidants.

BLOOMS: April to October.

BOTANICAL NAME: *Ziziphus spina-christi*

COMMON NAMES: The Sidr tree, lote tree, jujube, nabbag or nabkh tree. (It is believed that Jesus's crown of thorns was made from Sidr branches.)

PROVENANCE: Native to Sudan and the mountains of Hadramaut, Yemen, on the southwestern Arabian peninsula.

TERRIOR: Tropical to subtropical valleys and desert regions. Requires full sun and is drought tolerant.

HONEY COLOR: Very dark amber, like motor oil.

TASTING NOTES: Apple. Rich and buttery. Thick body.

PAIRINGS: Sidr honey is known to have many medicinal benefits.

63. SOURWOOD

PLANT CHARACTERISTICS: A small tree with white bell-shaped, droopy flowers that resemble lily of the valley. The fruit is a small, woody capsule and has a sour aroma, which gives the plant its name. It is an important honey plant of the South, and the honeycombs are so delicate that extraction is difficult.

BLOOMS: Late June through July.

BOTANICAL NAME: *Oxydendrum arboreum*

COMMON NAMES: Sourwood, sorrel tree, lily of the valley tree, elk tree.

PROVENANCE: The Appalachian Mountains of North Carolina, Tennessee, Virginia, Georgia, and Alabama.

TERROIR: Dry, acidic soil. Tolerates full or partial sun.

HONEY COLOR: Medium amber with a tint of light straw yellow.

TASTING NOTES: Complex, with hints of anise and gingerbread, and a warm finish. Delicate with just a touch of sour cutting the sweet taste. Sweet, spicy, anise aroma and flavor. A pleasant, lingering aftertaste. Does not crystallize.

PAIRINGS: Spread on cinnamon-raisin toast. Add to green tea, lemonade, coconut curry soup, or carrot soup and plum compotes. Add to glazes for pork chop or ginger chicken.

64. STAR THISTLE

PLANT CHARACTERISTICS: A thorny weed with yellow bushy flowers surrounded by nasty sharp spikes that protect the nectar and yellow pollen. An important honey plant.

BLOOMS: First of July to late August.

BOTANICAL NAME: *Centaurea solstitialis*

COMMON NAMES: Yellow star thistle, golden star thistle, yellow cockspur, St. Barnaby's thistle, *cardo, chardon.*

PROVENANCE: Native to the Mediterranean region. Found in Onoma, Napa, Solano, and Sutter counties in California, as well as Idaho, Michigan, Arizona, Oregon, and Washington.

TERRIOR: Dry, drought-tolerant soil. Poor soil on hedgerows, on barren hills, and in neglected fields.

HONEY COLOR: White or extra light amber with a greenish cast similar to that of olive oil.

TASTING NOTES: Wet grassy, musky, spicy, anise, and cinnamon aroma. Very sweet flavor. Heavy body. Rapid granulation. Cloying yet buttery.

PAIRINGS: Drizzle over strong cheeses and toasted pine nuts, and serve with corn bread.

65. SUNFLOWER

PLANT CHARACTERISTICS: A tall annual herb bearing hairy stalks of daisylike flowers. The face of the flower is brown, and the petals are yellow. These flower heads follow the direction of the sun, rotating from east to west during the day.

BLOOMS: July and August.

BOTANICAL NAME: *Helianthus annuus*

COMMON NAMES: Marigold of Peru, *corona solis, sola indianus, chrysanthemum peruvianum, girasole, tournesol, girasol.*

PROVENANCE: Grows in Georgia, Delaware, Minnesota, North and South Dakota, Nebraska, Wisconsin, and Utah as well as Italy, Austria, France, and Spain.

TERRIOR: Wastelands and prairies.

HONEY COLOR: Pale yellow to amber, turning dark yellow when crystallized.

TASTING NOTES: Can be bitter with nutty and ripe apricot overtones and herbaceous, citrus notes. Crystallizes rapidly and usually into a very compact mass.

PAIRINGS: Drizzle over yogurt and serve with sliced fresh peaches or nectarines. Spread on biscuits or rye sesame shortbreads. Used in making candy nougat.

66. TAHONAL

PLANT CHARACTERISTICS: Yellow daisylike flowers appear at the tips of long, slender, leafless stalks.

BLOOMS: January and February.

COMMON NAMES: Tajonal, goldeneye, plateau goldeneye, sunflower goldeneye, toothleaf goldeneye.

BOTANICAL NAME: *Viguiera dentata*

PROVENANCE: Yucatan Peninsula, Mexico. Found in Arizona, Texas, and New Mexico.

TERRIOR: An extremely drought-tolerant plant. Tolerates full or partial sun.

HONEY COLOR: Extra light amber to dark amber.

TASTING NOTES: Deep and tropical character. Wild and delicate flavor. Crystallizes rapidly.

PAIRINGS: This honey is used as a blending or baking honey. It must be used fairly quickly because of its high moisture content and to avoid possible fermentation.

67. TAWARI

PLANT CHARACTERISTICS: A large evergreen with magnificent silky, white-yellow flowers overflowing with nectar.

BLOOMS: November to December (New Zealand summer).

BOTANICAL NAME: *Ixere brexiodes*

PROVENANCE: Native to the east coast of the North Island of New Zealand and to lowland forests in northern New Zealand.

TERRIOR: Lowland and subalpine forests.

HONEY COLOR: White-yellow with a golden tinge.

TASTING NOTES: Known as the beaujolais of honeys because it is best enjoyed young. Subtle, musky flavor with a frothy texture. Overtones of orange, butterscotch, and licorice.

PAIRINGS: Great for spreading over pancakes, waffles, and ice cream.

68. THYME

PLANT CHARACTERISTICS: A strongly flavored perennial herb with white, pink, or purple flowers growing in short clusters. The leaves are evergreen in most species.

BLOOMS: October to November.

BOTANICAL NAME: *Thymus vulgaris*

COMMON NAMES: Common or garden thyme, *timo, thym, tomillo*.

PROVENANCE: Greece, Italy, eastern Europe, New Zealand's South Island around the town of Alexandra.

TERRIOR: Dry slopes, rocky clay, poor soil.

HONEY COLOR: Caramel to dark golden amber or darker.

TASTING NOTES: Aromatic, floral, lemony, minty, pungent, strikingly sweet, and acidic. Dense body and almost foamy crystallization.

PAIRINGS: Drizzle over Gorgonzola cheese topped with fresh figs and pecans on crispy bread, and serve with chardonnay or Syrah. Mix into Greek yogurt with granola. Stir into herbal tea, and mix into lemon sorbet. Use to make lemon-pepper glaze for lamb, fish, and poultry.

69. TULIP POPLAR

PLANT CHARACTERISTICS: An ornamental tree with stunning, fragrant, bell-shaped yellow-green flowers that are orange on the inside. It has four lobed leaves that turn a blaze of gold each autumn.

BLOOMS: April to June.

BOTANICAL NAME: *Liriodendron tulipifera*

COMMON NAMES: Tulip tree, white wood, yellow poplar, tulip magnolia, American tulip tree, *tulipifero, tulipier, tulipero.*

PROVENANCE: Native to the middle eastern seaboard of North America. An important honey plant in Virginia, Tennessee, Kentucky, Georgia, and North and South Carolina.

TERRIOR: Sandy, dry conditions. Prefers sun, yet is shade tolerant.

HONEY COLOR: Dark amber with a very deep amber reddish tinge. Becomes darker with age.

TASTING NOTES: Rich, pleasant, sweet, flowery, and smoky with a hint of metal. Resembles molasses.

PAIRINGS: Drizzle over blue or stinky cheeses and sliced green apples and pecans, and serve with Syrah. Pour over vanilla ice cream or sliced fresh peaches. Spread on pancakes, waffles, bran muffins, gingerbread, and corn breads.

70. TUPELO

PLANT CHARACTERISTICS: A gum tree with clusters of greenish flowers that develop into soft, red berrylike fruits. Tupelo honey is rare and is considered one of the most delicious honeys. In order to harvest the honey, honeybee hives are placed on platforms in the swamps where the tupelo gum grows. The process is expensive and labor intensive, making this honey sought after and highly respected.

BLOOMS: April and May.

BOTANICAL NAME: *Nyssa ogeche* (The name of the genus is derived from *Nysseides*, the name of a Greek water nymph.)

COMMON NAMES: Tupelo, nisa, Ogeechee tupelo. (*Tupelo* comes from two Cree words that mean "tree of the swamp.")

PROVENANCE: Native along the Apalachicola, Choctahatchee, and Ochlockonee rivers of Georgia and along the Chipola and Apalachicola rivers of northwest Florida.

TERRIOR: Prefers moist, well-drained, acidic soils of the pineland swamps. Likes full sun, but tolerates light shade.

HONEY COLOR: White or extra light amber with a greenish cast.

TASTING NOTES: Rich buttery texture. Floral aroma with herbal notes and with hints of cinnamon, melon, and pears. Because of its high fructose levels, the honey is very sweet and won't crystallize.

PAIRINGS: Drizzle over blue, aged pecorino, and other robust cheeses, and serve with cabernet sauvignon or Syrah. Mix as a glaze for pork chops.

71. ULMO

PLANT CHARACTERISTICS: A slow-growing evergreen shrub that has large white, camellialike flowers. Bees love its aromatic nectar. Ulmo trees are a threatened species of rain forest flora.

BLOOMS: January to March (Chilean late summer).

BOTANICAL NAME: *Eucryphia cordifolia*

COMMON NAMES: *Gnulgu, muermo, roble de Chile.*

PROVENANCE: Native to Patagonia, Chile, and the Cochamó Valley in Argentina. It is found in Araucania and Chiloe in Chile. Ulmo also grows well in Scotland and has been introduced in the north Pacific coast of the United States.

TERRIOR: Temperate rain forests along the Andes Mountains. This land is rich in humus, and the climate is humid.

HONEY COLOR: Light amber with pink tones.

TASTING NOTES: Creamy, buttery, with exotic perfume of aniseed, jasmine, vanilla, violet, and cloves. Touches of tea and caramel.

PAIRINGS: Drizzle over traditional Chilean chanco, panquehue, and quesillo cheeses. Pour over vanilla ice cream, dulce de leche, rice pudding, flan, and corn cakes. Use in savory sauces for fish.

72. VIPERS BUGLOSS

PLANT CHARACTERISTICS: A tall perennial wildflower that covers the hills of New Zealand's central South Island in a sea of brilliant blue. It has rough, hairy leaves, and the seed, resembling a viper's head, was once mistakenly used as a treatment for snakebites, thus giving the plant its name.

BLOOMS: December to March (New Zealand summer).

BOTANICAL NAME: *Echium vulgare*

COMMON NAMES: Borage, blue borage (not to be confused with the herb borage), blue weed, *mel de soagem.*

PROVENANCE: Marlborough and Otago provinces in New Zealand's South Island and Alentejo in the southeastern region of Portugal.

TERRIOR: Well-drained chalky or limestone soils in mountain valleys with hot, dry summers.

HONEY COLOR: White with a buttery brown tint.

TASTING NOTES: Delicate flowery bouquet. Hints of vanilla and bee pollen. Often very dry with a chewy or tacky texture. Slow to crystallize.

PAIRINGS: Drizzle over amarelo de beira baixa cheese, or mix with cocoa or coffee drinks. Served cold it is a favorite chewy snack for children.

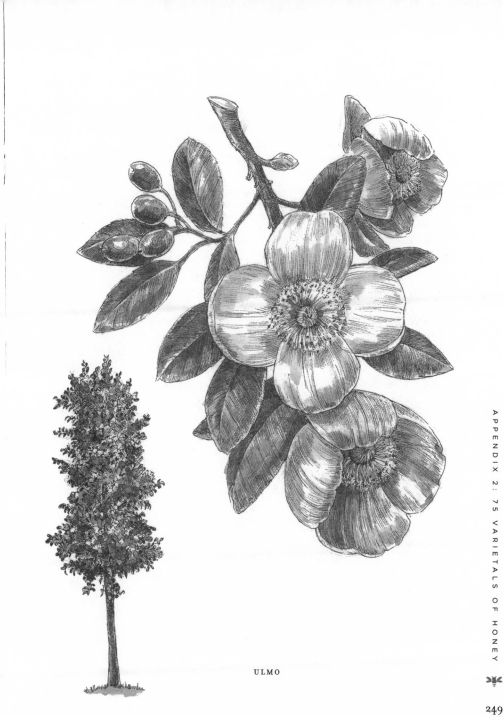

ULMO

73. WENCHI

PLANT CHARACTERISTICS: A tree with light orange and pinkish red flowers. In order for Wenchi honey to be harvested, bees are kept in traditional hives consisting of large cylinders made of woven bamboo and banana leaves. The hives are closed at one end by pieces of wood. The other end, where the bees enter, is closed with straw. The local bees are very aggressive, and beekeepers collect the honey during the night, using a lot of smoke to drive the bees away. This harvesting technique gives the honey a smoky flavor.

BLOOMS: At the end of the rainy season between November and January.

BOTANICAL NAME: *Hagenia abyssinica*

COMMON NAMES: African redwood, *brayera, cusso, hagenia, kousso*. Locally called *kosso*.

PROVENANCE: Native to Ethiopia.

TERRIOR: High-elevation, rocky, volcanic soil. Arid, hot climate.

HONEY COLOR: Yellow-amber with a reddish tinge.

TASTING NOTES: Intense flavor with a floral fragrance and notes of lightly roasted caramel and balsamic. Creamy on the palate, it has medium acidity.

PAIRINGS: Drizzle over sesame-seed bread and spiced breads. Used to make red-pepper spice paste that is an ingredient in lentil dishes. Used in marinades for chicken and slow-simmered meats. Used to make mead or honey wine.

74. WILELAIKI (pronounced *Willie-LIE-key*)

PLANT CHARACTERISTICS: Sweet-smelling, white, aromatic flowers followed by tiny red berries that glow against the dark green foliage.

BLOOMS: August to November.

BOTANICAL NAME: *Schinus terebinthifolius*

COMMON NAME: Mexican pepper, Florida holly, Christmas berry (in the United States), Christmas bush, pepper tree, Brazilian pepper, *poivrier sauvage*.

PROVENANCE: Hawaii.

TERRIOR: Tropical conditions. Cliffs, sandy soil, full sun.

HONEY COLOR: Light to medium amber, with a greenish tint.

TASTING NOTES: Smooth texture. Exotic and spicy with floral, smoky, peppery flavors. Hints of marzipan and chocolate.

PAIRINGS: A bakers-grade honey, used mostly in prepared foods. A good sweetener for coffee and chai tea. Drizzle over steamed carrots and almond cake or add to marinades.

75. ZAMBEZI (pronounced *zăm-bē'zē*)

PLANT CHARACTERISTICS: A leguminous forest tree. It sheds its leaves to conserve water during the overbearing hot season. Its red blossoms are a sign that the rainy season is beginning. This honey is certified organic, kosher, and fair traded.

BOTANICAL NAME: *Brachystegia longifolia*

COMMON NAMES: Miombo. Called *mutundu* in the local Swahili language.

PROVENANCE: Miombo forests of Zambia called *Ikelen'ge Pedicle*, considered one of the world's last ecologically diverse regions with virgin forest.

TERRIOR: Tropical to subtropical woodlands and savannas. Humid climate with long dry seasons. Poor soils.

HONEY COLOR: Dark amber, chocolate.

TASTING NOTES: Rich, smoky, spicy, and woodsy flavor with notes of toasted caramel and tobacco and the tang of red currants.

PAIRINGS: Used in traditional African cuisine made with collard greens, pumpkin seeds, chilis, and meat stews and wine making.

Glossary

ANTHER: The pollen-bearing part of the stamen, or male reproductive organ, of a flower.

APIARIST: A beekeeper.

APIOLOGY: The scientific study of honeybees.

APIARY: A yard or field where beehives are kept. Also called a bee yard.

ARTISANAL HONEY: Honey produced by individuals using traditional methods and thus preserving the integrity of the product. With artisanal honey, quality and character are highlighted, rather than quantity and consistency.

BEEHIVE: The man-made boxes where honeybees are raised. In nature, it can be any place where honeybees live.

BEE SPACE: The space needed for honeybees to move between frames within a hive. This space has been determined as being 3/8 inch.

BEESWAX: The wax that is secreted from the glands of the female honeybee's abdomen and then molded to make honeycomb.

BROOD: The immature stages of a honeybee's life: eggs, larvae, pupae.

COLONY: The social structure of honeybees. The colony includes workers, drones, and one queen.

COMB HONEY: Honey taken out of the beehive in its original state still inside the beeswax.

DRONE: The male member of the honeybee colony. Drones comprise roughly 10 percent of a colony's total population.

EXTRACTED HONEY: Liquid honey that has been separated from the beeswax by an extractor or spinner and then usually poured into jars.

EXTRACTOR: A large barrel-like machine that separates honey from the frames through centrifugal force. Extractors can be electric or manual. Also known as a spinner.

HONEYCOMB: Beeswax shaped into hexagonal cells made by honeybees and used for raising brood and storing honey and pollen.

LARVA: The newly hatched wingless form of the honeybee. The second stage of brood development.

MEAD: A fermented beverage made with honey.

MELISSOPALYNOLOGY: The study of honey by identifying its pollen sources.

NECTAR: A sweet liquid secreted by flowers, which is gathered by honeybees and made into honey.

POLLEN: A protein-rich, powder-like substance produced by the anther of plants and containing the male reproductive cells. Honeybees gather pollen as food for themselves and their young.

POLLINATION: The transfer of pollen from the anther to the stigma, often by the wind or insects, especially honeybees; the necessary step before fertilization of a flowering plant.

PROPOLIS: A sticky, waxy resin collected by honeybees from the buds of trees. Used by honeybees to close up openings within a hive that are less than approximately ³/₈ inch wide (the width of bee space).

PUPA: The non-feeding stage in the birth cycle of the honeybee, during which the larva develops into an adult. The last stage of brood development.

STIGMA: This tip of the pistil, or female reproductive organ, of a flower on which pollen is deposited during pollination.

SWARM: A group of approximately half of a colony of honeybees that, along with the queen, has left the hive to establish a new colony.

QUEEN: The only sexually developed female in a colony of honeybees. She is the mother of all the bees in the hive. A healthy hive can have only one queen.

VARIETAL HONEY: Honey that has at least fifty-one percent of its nectar from one floral source. Also called uni-floral or single varietal honey.

WORKER: Sexually undeveloped female bee that performs all the chores within a hive except laying eggs (during normal circumstances).

Bibliography

BOOKS

Beck, M.D., Bodog F. and Dorée Smedley. *Honey and Your Health*. New York: Bantam Books, 1971

Boardman M.D, Joseph. *Bee Venom: The Natural Curative for Arthritis and Rheumatism*. G.P. Putnam and Sons, 1962

Brown, Royden. *Bee Hive Product Bible*. New York: Avery Publishing Group, 1993

Caron, Dewey M. *Honey Bee Biology and Beekeeping*. Connecticut: Wicwas Press, 1999

Crane, Eva. *A Book of Honey*. New York: Charles Scribner's Sons, 1980

Crane, Eva; Penelope Walker and Rosemary Day. *Directory of Important World Honey Sources*. Great Britain: International Bee Research Association, 1984

Crane, Eva, edited by. *Honey: A Comprehensive Survey*. London: Heineman, 1975

Crane, Eva and Penelope Walker. *Pollination Directory for World Crops*. Great Britain: International Bee Research Association, 1984

Edwards, John. *The Roman Cookery of Apicius*. Washington: Hartley & Marks, 1984

Lovell. John. *Honey Plants of North America*. Ohio: A.I.Root, 1926

MacNeil, Karen. *The Wine Bible*. New York: Workman, 2001

Maurice Hanssen. *The Healing Power of Pollen and Other Products from the Beehive*. Great Britain: Thorsons Publishers, 1979

Modesti, Claudio. *I mieli uniflorali incontrano I formaggi traditionali italiani*. Italia: Edito dal Grupo Editorale Geronimo, 2005 (Single-floral source honeys and Italian cheese pairings)

Morse, Roger A. and Mary Lou. *Honey Shows*. Connecticut: Wicwas Press, 1996

Murat, Dr. Felix. *Les Trois Aliments Miracles and Le Pollen-Alin Caillas*: 1971 Bee Pollen Miracle Food (Brochure)

Nixon, Gilbert. *The World of Bees*. Great Britain: Arrow Books LTD, 1959

Root, Amos Ives. *The ABC & XYZ of Bee Culture-41st Edition*. Dr. Hachiro Shimanuki, Kim Flottum, Ann Harman, 2006

MAGAZINES

American Bee Journal

Apitalia

Bee Craft

Bee Culture Magazine

Journal of the American Apitherapy Society